The Stable Framework™

Michael J. Berry

www.StableFramework.org

ISBN-13: 978-0692144008
ISBN-10: 0692144005

Author: Michael J. Berry
Publisher: Stable Framework Publishers
Location: Salt Lake City, UT   USA

www.StableFramework.org

First Edition
1.19

ACKNOWLEDGMENT

I acknowledge and thank Dr. Bret Swan for his encouragement while creating The Stable Framework™.

Forward

Michael Berry's book entitled *The Stable Framework* is an excellent read for organizations in the IT industry that desire to evolve quicker than their competitors and ultimately become the future dominators within their industry.

Often, and for good reason, the IT industry is compared to the manufacturing industry in hopes that IT can use traditional manufacturing tools to evolve at a greater pace. What people don't usually understand is that the manufacturing industry started out with a 2% production success rate and that it took 220 years to get from a 2% to a 99.99964% success rate, which is a sustainable capability for a modern manufacturer. Another interesting fact is that Eli Whitney, the first American manufacturer, missed his first contracted delivery date for 12,000 muskets to the US Government by 12 years.

In time, manufacturers started to understand the need for generating organizational cultures that support analytical tools. If organizational culture does not support analytical tools, then the analytical tools do not work. Hence, for an organization to be capable of effectively using analytical tools to generate profound knowledge about how to improve its capabilities, it needs to have a supportive culture in place.

If the IT industry evolves at the same rate as the manufacturing industry, it will reach similar success rates around 2188 which is 220 years after John Tukey first published the term "Software." Of course, the hope continues to be that if we can transfer lessons learned from manufacturing to IT this evolution will go much quicker—and I believe it will—however, it is not as easy as one may think to transfer these lessons.

Many books try to teach the IT industry how to use heavy-duty analytical principles even though most IT companies are not yet ready for them. In other words, before their processes are predictable. What makes Michael's book unique is that he does not introduce these highly analytical tools, as he recognizes that the IT industry must first achieve process predictability before advanced mathematics can apply. The Stable Framework™ is a book written to address the current needs of the IT industry. Michael brings out the tools that were and are used in the manufacturing industry to create predictability. If IT organizations will effectively implement this knowledge into their organizations, they will have put themselves on a course to become the powerhouses of the future.

I consider Michael to be a visionary to what the IT industry must become and what it is going to take to get there.

—John L. Lee
  2016 Shingo Research and Professional Publication award winning author of:
  Rising Above It All: The Art and Science of Organizational Transformation

Table of Contents

Table of Figures

Introduction

Deming warned us. We are in a crisis.

The first interchangeable part design in recorded history was attributed to Jean-Baptiste Vaquette de Gribeauval, the French cannon maker. In the 1700's he helped Napoleon win many battles by creating lighter-weight cannons that were faster to produce and more resilient because they had interchangeable parts. Napoleon may have been a strategy genius, but having better cannons was an enormous tactical advantage.

Inspired by this innovation, Honoré Blanc demonstrated that he could create muskets with interchangeable parts in the late 1700's. Afraid of upsetting the labor guilds, government leaders in Europe were skeptical and unreceptive to investing in Blanc's concept. An American ambassador to France named Thomas Jefferson was the exception. He realized this capability would free America from dependence on European governments for military supplies.

Jefferson tried to convince Blanc to move to America and further develop his concept there, but Blanc was unwilling. Jefferson eventually returned to the states and convinced President George Washington to fund the idea. In 1798 a contract was issued to Eli Whitney for 12,000 muskets. Whitney was known for inventing and manufacturing the Cotton Gin which transformed the southern states. Although Eli Whitney produced all the muskets ordered, they were delivered 12 years late.

In 1855, Samuel Colt, inspired by a steam ship's paddle-wheel, invented the 6-shooter revolver and designed a factory with the help of Eli Whitney's son to produce the guns using interchangeable parts and a new idea called an assembly line. With this combination, Colt was able to produce 5000 revolvers a year, giving a huge advantage to the Texas Rangers and the American Army, enabling them to win the Mexican-American war.

Samuel Colt's factory caught the attention of Henry Ford, who copied those ideas and mass-produced the Model T automobile. Henry Ford's factory employment model—offering a steady job with an eight-hour workday—was the beginning of the middle class in America.

The "American System of Manufacturing" quickly became a worldwide standard. Industrialists around the world raced to duplicate the concept of using an assembly line and interchangeable parts to mass produce goods.

In 1911 Frederick Taylor introduced the concept of "Scientific Management," demonstrating that pre-planned work tasks on assembly lines can be scientifically optimized and tuned to match each worker's "Personal Coefficient" to enable sustainable, optimized production. This combination benefited both the worker and the owner because the operation could be tuned to produce the maximum possible output and the best quality of work for the smallest combined costs of human effort, natural resources, and other capital expenses. As management and employees worked together to widen the gap between top-line income versus bottom-line expenses, the organization could afford to pay the employees better wages. If they collaborated, they would both win.

Although production yield increased, quality management at that time was in its infancy. Inspectors waited at the end of assembly lines to separate non-conforming goods from sellable goods. Non-conforming goods were set aside for someone to examine and then rework, discard, or occasionally forward on to an unlucky customer.

In 1924, Walter Shewhart introduced the "Statistical Process Control Chart" while working for a supplier to Bell Laboratories. Using process data sampling, Shewhart demonstrated that every process contains natural variation and that it is possible to measure this variation and detect if a failure in the process was caused by inherent randomness, or by an external assignable cause which had negatively impacted the process. This was important because the approach used for addressing either type of failure should be different. Shewhart also defined the famous "Plan-Do-Check-Act" model, which is used to understand a system better so that the variation within a system could be identified and minimized. In addition, he worked with a fellow statistician, Joseph Juran, to develop an idea called TQM, or Total Quality Management. This is the idea that quality is the responsibility of everyone—not just the inspector at the end of the assembly line.

Several years later in 1927, a young statistician named W. Edwards Deming met Walter Shewhart and began applying his ideas of statistical process control beyond manufacturing and into management. He introduced the concept of "Systems Thinking," where any change made to a component of a system should first be examined for any impacts it may have on the entire system, and potentially on the suppliers and consumers of the system. Deming further proposed that people were just one of many variable factors within a system and that while the employees should be responsible for performing a job correctly, management should be responsible for minimizing variation within the system.

Deming approached the major U.S. automobile manufacturers with his innovative concepts but found no interest. The prominent management thinking of the day was to find and hire good workers and then incentivize or threaten them in order to meet quotas. Why should they listen to his ideas which pointed blame at management for most production problems? The "get it done" quota mentality took over in America and we stopped progressing with quality innovations.

At this point, WWII started and ended. The great industrial nations of the world were devastated, except for the United States. Immediately after the war, American manufacturers were able to export and sell almost everything they could build. Quality was becoming much less important and was certainly not a primary concern. In fact, as the quota mentality grew, an attitude started to brew in America that "quality kills."

In 1949, the U.S. Army asked Deming to assist with the census work being conducted in Japan. While there, Deming gave a landmark speech in August 1950 to a room full of Japanese industrialists about Systems Thinking, Statistical Process Control, and Total Quality Management. He was soon asked by many of the manufacturers to stay in Japan and help them rebuild their factories using those ideas. He did.

Months before this landmark speech, the U.S. Army sent Eiji Toyota, a senior engineer at his cousins' company, to visit Henry Ford's automobile plant in Dearborn, Michigan. The Army wanted the Toyota Motor Company to build trucks for the U.S. troops in Korea because it was cheaper to ship them from Japan than from Michigan. Eiji Toyota returned in awe. While Toyota had produced only 2,500 cars in its 13-year history, Ford was manufacturing 8,000 cars a day. Over the next few decades, Eiji Toyota worked with fellow engineer Taiichi Ohno to use mass production ideas from the Ford plant, but they adjusted them with a focus on quality. Together, they created the foundation of the Toyota Production System (TPS).

TPS evolved to include a list of eight sources of waste, a streamlined "Just-In-Time" supplier delivery system, and innovations such as

Kanban cards, kaizen quality circles, and inventory-pull systems. Deming and Taiichi Ohno eventually became acquainted through the Japanese Union of Scientists and Engineers (JUSE).

During this time, Taiichi Ohno discovered that having respect for coworkers and regard for the individual laborer was a necessary attribute in a quality system. His philosophy was that people should be respected and encouraged to contribute and grow. After all, how can people continuously improve an environment if they are not continuously improving themselves?

Toyota hired an independent consultant named Shigeo Shingo to teach these concepts to thousands of Toyota employees in 1955. He later had several of his training books translated into English and became a fundamental catalyst in introducing the West to Asian manufacturing techniques that we know today as the Toyota Production System (TPS).

Shigeo Shingo went on to apply these principles at Mitsubishi which reduced the time needed to produce a supertanker from four months down to two, setting a world record. A major innovation Shingo developed and used extensively was a source inspection technique called the "Poka-Yoke." This is a tool developed to ensure a product, component, or step is performed correctly. Using an innovative tool like this ensured zero defects and eliminated most of the need for statistical process control.

In 1962 Kaoru Ishikawa introduced the concept of Quality Circles to JUSE. Quality Circles are activities where workers that do similar work meet regularly in small groups to identify, analyze, and solve work-related problems. Today we call these "Kaizen Circles."

Kaoru Ishikawa later introduced the Ishikawa Diagram, in 1982, for performing root cause analyses of challenging problems experienced during production.

While Deming stayed in Japan, helping various companies improve their manufacturing processes, technologies continued to advance in America.

Philip Crosby became the senior quality engineer for the Pershing missile manufacturing process in the 1960s. His experience there led to the publication of a popular book in 1979 called *Quality is Free*, where he emphasized the DIRFT concept, or "Doing It Right The First Time." This concept emphasized preventative-based quality, a zero-defects goal, a "price of nonconformance" quality measurement standard, and a definition of quality being "conformance to product and customer requirements."

In 1984, Eli Goldratt published his famous book *The Goal*, where he introduced the "Theory of Constraints." This is a concept that emphasizes the importance of prioritizing the flow of value through a manufacturing environment and suggests that companies should continually identify and remove any bottlenecks found within a process. This flow of value, called a "value-stream," has become another focal point for efficiency. Companies are now re-structuring their entire operations around this concept. Instead of making activities and assets easy to manage which tends to encumber the flow of value, they are making the smooth flow of value the priority, which requires advanced planning to manage the activities and assets.

In 1986, Alfie Kohn published a compelling book called *No Contest,* expressing the enormous benefits of people cooperating with each other instead of competing. He pointed out that our Western American culture is addicted to competition, and that we would accomplish much more together if we learned to cooperate rather than compete. Deming became a big fan and advocate of Alfie Kohn's ideas.

That same year William B. Smith, Jr. and Dr. Mikel J. Harry developed a set of tools at Motorola, based on Shewhart's Statistical Process Control techniques, for identifying and minimizing variations in a process. They called their new program Six Sigma. It was soon adopted as a core business strategy by Motorola and then General Electric before going mainstream.

In 1988, a young MIT student named John Krafcik wrote a paper about his observations made while visiting several factories in Japan. He observed that all of the workers actively engaged in quality improvement activities. In his paper, he called what he observed "Lean." This name has now become the western version of the practices and philosophies derived from the Toyota Production System.

Although the Western outlook on quality was maturing, Japan had a 30-year head start, and a culture better tuned for group work. As Japan joined the world economy, the crisis Deming would soon warn us about began.

The first major victim of the quality revolution was the U.S. electronics industry, which died in the 70s. American electronics manufacturers such as RCA and Zenith lost their businesses to Sony, LG, Panasonic and other Asian companies. Nobody seemed to ask why. American electronics manufacturers simply assumed the Asians were just willing to work for lower wages, which lowered the price of their well-built products.

The second major victim of the quality revolution was the U.S. auto industry, which in the 1980s realized its products were statistically inferior to Asian competitors and were now starting to lose market share to Asian companies.

By 1982, the Ford Motor Company was in full crisis mode. The company had accumulated $3 billion in losses during the previous four years and management knew they needed to change something. In desperation, Ford sought out Deming and asked for his help. Thanks to a timely NBC television broadcast that aired in June of 1980 called "If Japan Can, Why Can't We," Deming publicly showcased the improvements that revolutionized Japan and challenged American companies to follow suit. He spent the next few years helping Ford re-educate its management teams on ways to create a culture of quality, and practice proven quality principles. By 1986, Ford had become the most profitable car company in America, beating out General Motors and Chrysler for the first time since 1920.

That same year Deming published his famous book, *Out of the Crisis*, in which he warned modern industrial nations that we are now in a crisis. The landscape has changed. No longer will mediocre quality levels maintained to fulfill production quotas keep a company profitable in a world-market. No longer will punctuality, work ethic, and following directions be all that is required to keep a company competitive.

In addition to encouraging a culture reset in favor of teamwork, employee empowerment and continuous improvement, Deming warned that the solution to progress is not expensive robots and computer technology. Those systems add costs to the final products. Anybody can build a better product eventually using expensive technology. The trick is to build a better product for less cost, not more.

On January 28, 1986 the Space Shuttle Challenger exploded 73 seconds into flight, on live television, in front of the entire world. Killing all seven crew-members, it was the worst accident in the history of the U.S. Space Program.

The next year the rest of the world got serious about quality. In 1987 a series of quality initiatives emerged. ISO 9000 was published by the International Organization for Standardization. The Malcolm Baldridge award was created by the U.S. Government. The Software Engineering Institute (SEI) started developing a software development quality standard called the Capability Maturity Model integrated (CMMi).

Two years later the British Government published the Information Technology Infrastructure Library, or ITIL, and COBIT was released a decade later in 1996.

Most of these programs originated from government entities. They were created primarily for internal use, military, and government vendors. Although they have gained popularity in some industries, they were designed to ensure quality and thoroughness, rather than value flow, or production speed. A common criticism for all of them is the enormous amount of process additives and additional personnel required for a reasonable implementation. In addition, certification costs are prohibitive for all but the largest companies.

So, like the early U.S. automobile manufacturers, quality practices in IT today still closely match the "find and hire good workers and then incentivize them" model.

To be world-class today, a company must be trained on basic quality concepts and actively practice them. A company must elevate quality initiatives above quota goals and relentlessly improve relationships with its suppliers and its customers. A company must understand Systems Theory and learn to always be improving its systems. A company must understand and systematize its repeatable processes and continually try out new ways to remove waste and minimize defects and variation. It must cultivate a culture from the top that supports collaboration, teamwork, and trial and error to achieve quality. All of this requires dedication, persistence, and support from senior management. It also requires the fundamental understanding that certain practices performed up front to ensure downstream predictability are a better trade for time than not embracing those practices and being left a victim of the unpredictable problems that ensue, and the unbudgeted time disruptions required to correct them.

I'm writing this book because many of the same quality principles used in manufacturing are directly applicable to the IT Industry. Some people even describe IT as "assembly lines of decisions." Think of each assembly line as a system. There are systems with repeatable processes in Operations, DevOps, Implementation and Development. Each of these systems contain inputs, throughputs, and outputs. Included in these systems are customers and suppliers that need engagement, and management teams that need to grow their businesses.

When the software industry began it had no structure and no best practices. In time, object-oriented coding techniques and tiered-architectures were developed so that our apps would scale as software development expanded to become enterprise worthy.

As time went on, Rapid Application Development techniques were created to enable us to deliver value in portions to a growing competitive marketplace using iterative and incremental models.

In 2001, the introduction of Agile techniques enabled better team interactions, better methods for staying focused, and a mandate to involve customers throughout the development process. This combination equipped Development teams with the focus and accuracy necessary to rapidly develop and deliver accepted value to the customer.

While Development efforts are creative experiences producing novel deliverables every release, support functions such as Operations, Implementation, and DevOps, are mostly repeatable efforts. The former is a journey in innovation, while the latter is a mandate to maintain or improve the existing customer experience while lowering costs over time.

The Stable Framework™ empowers all of these groups by combining Agile and Lean concepts in ways that provide relentless improvement in all areas over time. We call continuous improvement and the ability to display it Operational Excellence.

As you'll see, the Stable Framework™ is applicable anywhere a team performs repeatable processes. To every hard-working, sincere, and well-meaning IT group out there working in a reactive environment, following the practices in this book will change your life.

Chapter 1 - The Problem

In Dr. Atul Gawande's book, *The Checklist Manifesto,* we read about the tragic story of the first B-17 crash.

On October 30, 1935, the Boeing Company was demonstrating its newest large warplane to what is now the U.S. Air Force at Wright-Patterson Air Force Base, near Dayton, Ohio. The aircraft was a marvel of engineering. It was the first aircraft with four engines. The wings were so large they contained crawlspaces allowing the crew to access them during flight. The plane could carry heavy cargo and even had a kitchenette for long distances. The most striking feature was an array of armaments making it what some called a flying battleship, or fortress.

That Wednesday morning the plane was queued up to launch and demonstrate its majestic abilities in front of a set of Air Force senior observers. Onboard, piloting the craft was the Air Force senior test pilot, Major Ployer P. Hill, the Boeing Company's senior engineer, Les Tower, and several additional observers.

The plane took off and entered a steep climb. It headed almost straight up then tipped to one side. Stalling, it abruptly fell to the ground, crashing dramatically in front of the spectators and exploding in a ball of fire. Both Les Tower and Major Hill died that day.

The evidence submitted to the inspection board indicated that the root cause of the crash was that a gust lock mechanism protecting the flaps from moving around while on the ground had not been disengaged before taking off, making it impossible for the pilot to control the aircraft once it left the ground. Someone had forgotten to unfasten the lock before takeoff, and there was no process or additional mechanism in place for the pilot to check for this. In fact, at that time, pilots were supposed to get proper training once and then operate an aircraft thereafter from memory. A retrospective analysis by Boeing concluded that the number of distinct attributes in these newer aircraft designs had evolved to such a degree that there were just too many factors to manage from a pilot's memory alone. Boeing's solution was to create a pilot's checklist and require its use before takeoff on all future flights.

The B-17 Flying Fortress soon became a huge success, with a total of 12,726 units produced for the war. As a result of this catastrophe the pilot's checklist became a critical tool used to manage complexity for every major industry and is a primary component in most published quality frameworks.

Like a modern, complex aircraft, today's IT environments have a variety of complex components that must be coordinated correctly to produce value. Entire teams work together to collect, enrich, and handoff physical and intellectual assets from suppliers to customers in an ongoing stream of effort. Every step along the way has its own set of attribute data. Some simple. Some complex.

Sometimes the communication is good, sometimes it is not. Assumptions are made. Decisions are made. Some of them are correct. Some are not. Few, if any, processes are systematized, making an improvement only beneficial until someone forgets about it. All of this creates value, rework, and scrap. Customers pay for all three, but they would prefer not to pay for rework or scrap. We call rework, scrap, and unnecessary delays the "Hidden Factory" within a Production or Operations environment. This Hidden Factory is the problem the Stable Framework™ addresses.

The Stable Framework™ enables organizations to deliver quality products or services, for less cost and time, by reducing the Hidden Factory that was previously part of the system.

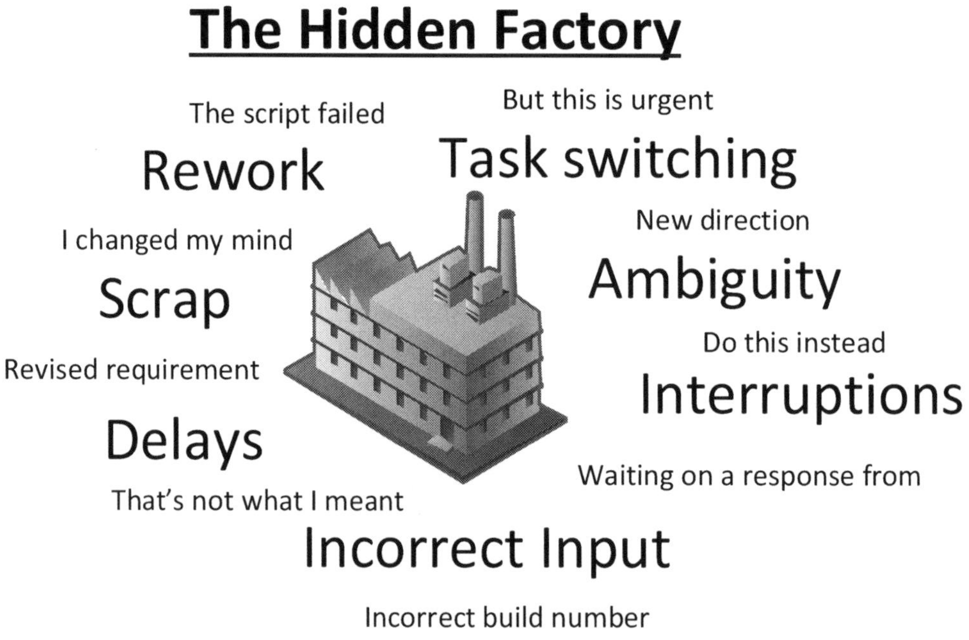

The Hidden Factory

The script failed

But this is urgent

Rework

Task switching

New direction

I changed my mind

Scrap

Ambiguity

Do this instead

Revised requirement

Interruptions

Delays

Waiting on a response from

That's not what I meant

Incorrect Input

Incorrect build number

Figure 1 - The Hidden Factory

Chapter 2 - The Hidden Factory in IT

The Standish Group International is a consulting firm that publishes a report card for the United States software industry every few years. With software being close to a $225 billion industry in the U.S., it's interesting to examine the size of its Hidden Factory. It's impossible to measure this number exactly, but the Standish Group's data provides some valuable insight.

One metric Standish publishes is the ratio of successful versus failed versus late projects. The firm defines a successful project as a project delivered on time, within budget, and containing all the targeted scope. Standish's data showed that only a small percentage of software projects met this definition. Also, a large percentage of projects were considered "failed," meaning they were completely scrapped.

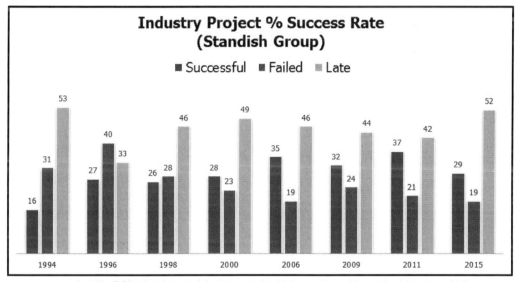

Sources, Standish Group CHAOS Reports: 1994, 1996, 1998, 2000, 2006, 2009, 2011, 2015

Figure 2 - U.S. Software Report (Standish Group)

This includes projects completed and deployed and promptly rejected by the customer. The data showed that during the past ten years this number was about 20% of all projects. In addition, this data showed that about 45% of the projects were significantly late.

While failed projects are usually rejected because the customer can't use the finished product, late projects are late because of the Hidden Factory.

The data shows that after Agile was introduced into the development community after 2001, project failure rates diminished about nine percent, and project success rates increased about nine percent, but the percentage of late projects hasn't change even one percentage point. In other words, Agile has help the industry produce more accurate software, but not an any more predictable manner.

Agile development patterns helped the industry lower project failure rates by involving the customer throughout the development process, instead of just the at beginning and the end.

The next evolution the industry needs is a Lean-based quality program like Stable to address the Hidden Factory. This solution would lower the costs of developing and supporting software and services by minimizing mistakes, scrap, and rework. A byproduct of this next evolution would be lower late project rates.

In my travels, I've had the occasional shock of consulting with a development team as they tell me they are on their second attempt at a multi-million-dollar project because the first attempt was scrapped. I ask them what are they doing differently this time to ensure success? They stare back at me like nobody had ever asked that question before, and they have no good answer.

Technical Debt

The term Technical Debt is sometimes confused with the term Hidden Factory, but they are not the same thing. Technical Debt is a concept in software development that reflects the implied cost of additional rework caused by choosing an easy or limited technical solution now instead of using a better approach that would take longer.

While Technical Debt doesn't need a resolution until an expansion is needed, the Hidden Factory must always be resolved.

Quality Debt

Often, project problems are not discovered until later in the production, implementation, or the service cycle. Studies conducted in 1999 and

2004 underscore the reality of the hidden cost of problems discovered later. The studies found that as defects are created in a workflow, they

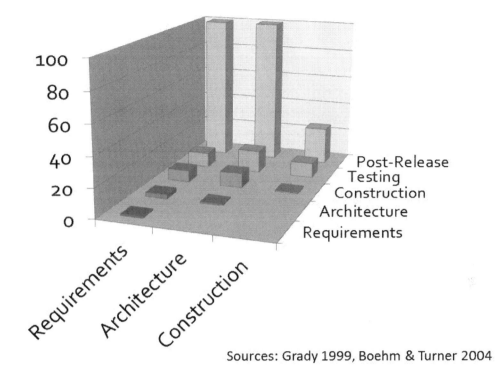

Sources: Grady 1999, Boehm & Turner 2004

Figure 3 - Exponential Cost of Fixing Software Defects

become exponentially more expensive to fix the further down the life-cycle they travel. This means that a defect created in the Requirements phase is relatively inexpensive to fix if found during that step in the process. However, once the team has coded the wrong requirement, the same defect becomes more expensive to fix. If that same problem requirement is found during testing, then more people are involved and the rework consumes more company resources. This effect, called "Quality Debt," is accumulated all the way down to where the final customer is impacted.

The purpose of a quality program is to create a framework that enables employees to find potential problems *within the same step in which they occur*, instead of falling back to "We'll deal with that problem later."

Quality Competence

The "Stages of Competency" model is a tool developed by Noel Burch in the 1970s. The model suggests awareness emerges in these four stages:

A. Individuals are initially unaware of how little they know, being unconscious of their incompetence.

B. As they begin to mature and become aware that tools and methodologies are available to help them achieve their goals, they recognize their incompetence and gradually become conscious of their incompetence.

C. As they begin working toward and showing improvement, they become consciously competent.

D. After much learning, experience, and success, they master the skill and reach unconscious competence...a Zen-like state.

Most organizations I visit are in need of a simple quality program. In the *Stages of Competency* learning model, they are in quadrant A or B. These companies are often unaware that patterns exist to transform their chaotic work environment into a stable work environment where they can operate with much less variation, and therefore, must less risk.

Stages of Competency

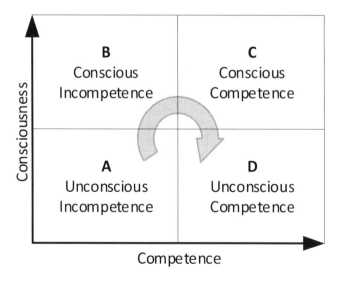

Figure 4 - Stages of Competency

Chaotic environments allow problems to grow undetected until they reach crisis mode and then require a hero to solve. Frequent heroic acts, as well as a recurring need for hot-fixes, are tell-tale signs of a chaotic environment.

Chaos can exist even among highly skilled people. Like the skilled Air Force test pilot, there is so much complexity in today's IT environments, that it is impossible for even a well-trained professional to repeatedly get everything right.

In contrast to a typical, rushed chaotic IT environment, picture your people looking at a performance console, reporting that there are no emergencies to fix. Picture them looking at performance indicator charts that show all service level agreements have been met and customer satisfaction trends are up. Picture them discussing remaining variances in their standard work results and working daily as a team to improve performance. Picture an orderly, clean environment where the team is learning and improving together. What you're envisioning is an out-of-the-box quality program called the Stable Framework™.

Chapter 3 - Systems Thinking and Value Streams

Customers want value. They want a solution to solve a problem they have, or to lighten a burden in their environment, or to improve their work and their lives. Businesses provide these solutions in the form of "Value Propositions." It's critical to understand that value is defined by the customer. It doesn't matter what the business thinks is valuable, value is defined entirely by their customers.

There are many attributes a customer considers when evaluating a value proposition from a potential supplier. Here are some common considerations:

Utility Value is a measure of how well the product or service matches the intended use by the customer. The safest way to ensure strong Utility Value is to involve the customer in the concept, design, and acceptance testing of a product or service. Deming taught us that it doesn't matter what we think is valuable, it's what the customer thinks is valuable that sells.

Warranty Value is the invisible trust associated with the seller of the product or service, and the customer's impression of the seller's ability to fix or replace that product or service upon failure. If the price is the same, it's always better to buy from a larger company or seller who will still be in business next year and can repair or replace the item for you.

Durability is a measure of how well a product or service withstands use. A proper quality program helps an organization build a robust or durable product and build it right the first time. Through continuous improvement the product or service is made to be more robust over time.

Price is what each customer evaluates, in terms of total cost, while contemplating whether to buy a product or service. Customers want to solve a problem without the risks or time involved with creating a solution themselves. In the end, they want to feel that what they purchased was more valuable to them than the money itself.

Accessibility means how easily a product or service can be acquired. Customers will pay more for a product or service that is easier to obtain and implement.

A quality program helps:

- Develop utility value by engaging customers to identify what basic needs and improvements are most urgent.

- Improve the durability and stability of any product or service offered by continuously improving past problem areas and developing and maintaining process controls.

- Reduce costs over time by minimizing rework, scrap, unwanted delays and unwanted variations in a product or service.

- Improve accessibility by increasing the predictability of release schedules as a result of minimizing variations within the system.

Systems Thinking

Deming taught executives to think in terms of the whole system and to understand how variation works within it. He taught that any change made to a component or process step potentially impacts the whole system it is in, and possibly the external systems interacting with it. He emphasized that the entire system must be understood before making changes to it.

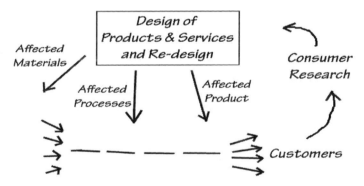

Figure 5 - W. Edwards Deming's System-Model

Deming taught about systems using the preceding diagram. He explained that a system has a trigger, input materials and suppliers, internal process steps, which add value to those materials producing an output product or service, and finally, the customers who consume the service.

A healthy system has a design and re-design component, which is always adding value based on continuous customer feedback, or lowering the cost and time required to produce that value through retrospection and innovation.

System Interactions

Systems interact with other systems. They often contain subsystems, and likely are a part of a supersystem. A domain of all systems considered is called an ecosystem.

An ATM system, for example, services a customer walking up to it and inserting his or her card to withdraw cash. The customer is the consumer system, and the bank computer containing the account information is a supplier system. The ATM is a system providing a service to connect the other two systems, any time of day. The PIN number authentication process could be thought of as a sub-system, and the entire bank corporation as a supersystem, or even an ecosystem.

It's important to realize a system can't understand itself. One must be outside of a system, talking with a customer to understand what the system needs to do. Fish in a fishbowl don't know they are in water. Humans thought the Earth was flat until they began studying the cosmos and triangulating external coordinates. You must get a perspective from outside of a system to understand it. This is a fundamental principle in system analysis and design.

The Two Vectors of System Excellence

Deming taught that a system must be continually improved. There are two vectors that govern system improvement. The first system improvement vector is how our system can adapt to better match the expectations of our customers. The only way to accomplish this is to communicate with them. Customer needs will evolve over time, so this communication and adjustment process should be ongoing.

After the customer's needs are being met, the second system improvement vector is to maintain the targeted level of service, but to deliver it faster, with less internal effort, which lowers costs over time.

To facilitate these vectors, consumer research from customer interactions must flow into a design and re-design system component, continually improving the output value for the customer, or reducing the effort, time, or cost to the business while working the system.

We call the first vector effectiveness, and the second vector efficiency.

Together, they represent the pursuit of excellence.

Mapping a System

To understand the components of a system, we must map the system. There are three approaches to mapping a system. The first is to use a S.I.P.O.C. Model to identify the components of a system. The S.I.P.O.C. Model shows the Suppliers, Inputs, Processes, Outputs, and Customers of a system. This can be useful for general discussions and evaluations.

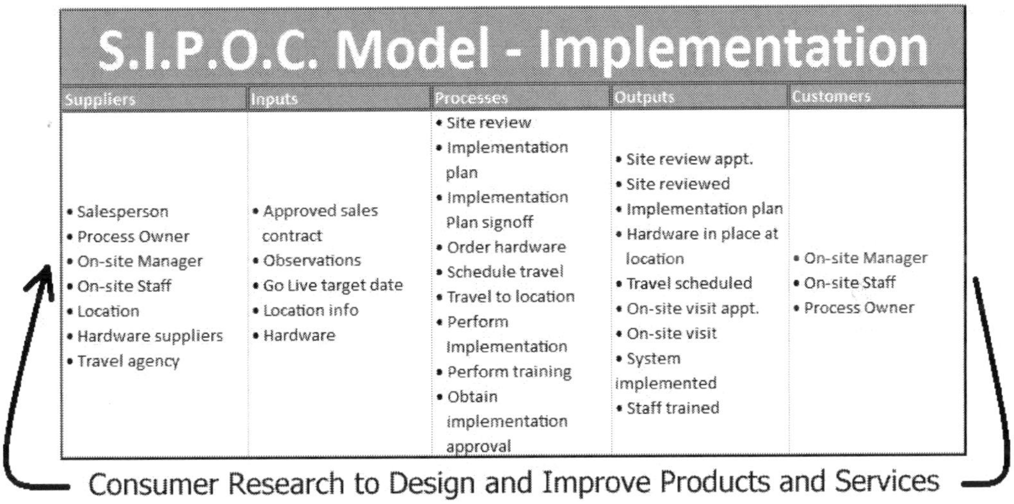

S.I.P.O.C. Model - Implementation

Suppliers	Inputs	Processes	Outputs	Customers
• Salesperson • Process Owner • On-site Manager • On-site Staff • Location • Hardware suppliers • Travel agency	• Approved sales contract • Observations • Go Live target date • Location info • Hardware	• Site review • Implementation plan • Implementation Plan signoff • Order hardware • Schedule travel • Travel to location • Perform Implementation • Perform training • Obtain implementation approval	• Site review appt. • Site reviewed • Implementation plan • Hardware in place at location • Travel scheduled • On-site visit appt. • On-site visit • System implemented • Staff trained	• On-site Manager • On-site Staff • Process Owner

Consumer Research to Design and Improve Products and Services

Figure 6 - The S.I.P.O.C. Model

You can describe an entire value stream using this diagram. In this case, we are describing an on-site software implementation process. It may be that this level of detail is enough for documenting a process or a system you are modeling.

System-Flow Diagrams

The second approach to mapping a system is by using a flowchart. This is useful if you need more detail to describe the asset movement, chain-of-custody, decision points, or division of labor within your system.

To create a flowchart, or "System-flow Diagram," we start by identifying the trigger. The trigger is the starting point for a process. Common examples of triggers are new sales orders, pull requests, new Service

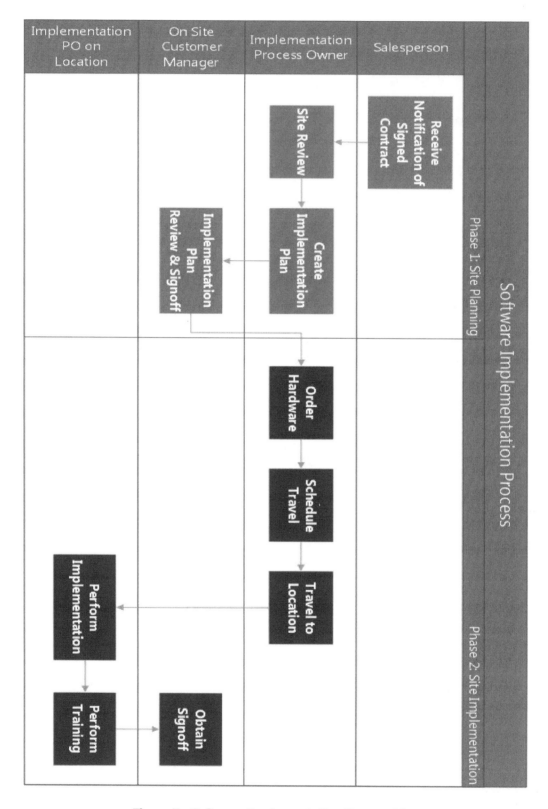

Figure 7 - Software Implementation Process Map

Desk tickets, end-of-business-day cycles, or end-of-month cycles.

Once we've identified the trigger, we then outline the process steps needed to fulfill the request. In the previous diagram you can see a Software Implementation Process Map. This map is organized by horizontal lines, called Swim Lanes, which indicate who is responsible for each step, and then vertical lines, which represent phases of the process.

In this example, there are two processes that occur in parallel to complete this value proposition in this system. The first process is triggered when the sales contract is signed. The second process is triggered when the on-site customer manager approves the implementation plan.

The value delivery for this system is complete when the on-site customer manager signs off the completed implementation.

Systematization

In order to be sure these processes are performed correctly, we'll establish Standard Operating Procedures and checklists for each process step.

A Standard Operating Procedure describes a process and should be as simple as possible, because IT tends to evolve quicker than other disciplines. It's important to indicate the necessary steps, but keep descriptions as brief as possible. The target reader should be someone skilled in the art, but new to the environment. Diagrams, simple bullet lists, SIPOC Models, or even photos or video instruction will work, too.

Think of checklists as certificates of proof the process was performed correctly. They are not instructional checklists so much as they are risk management checklists, indicating the required steps have been taken, and therefore all known risks have been checked before this work item can move on to the next step in the process.

Value-stream Maps

If you want more detail about where your system contains waste, a third option is to create a Value-stream Map. A Value-stream Map is like a

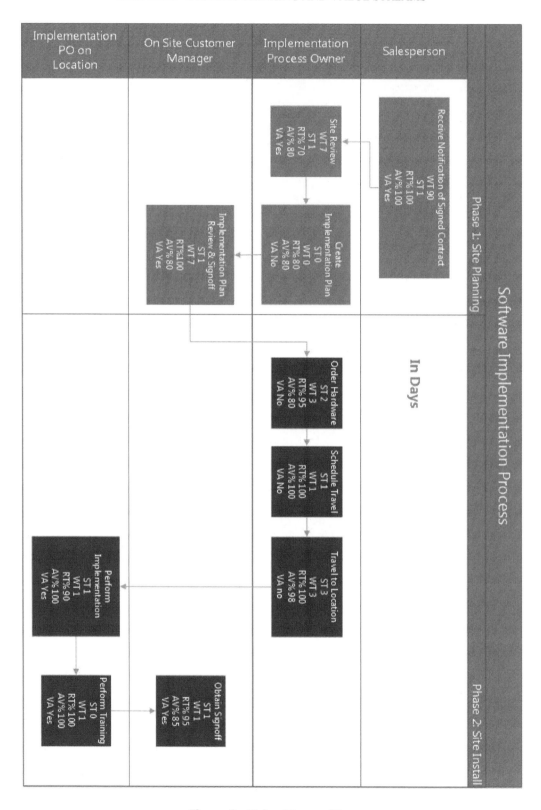

Figure 8 - Value Stream Map

System-flow Diagram except that each step contains additional timing, dependability, and value-add information.

For each step we'll assign the typical Step Time (ST), Wait Time (WT), Roll Through Percent (RT%), Availability Percent (AV%), and indicate if the step is considered a Value Add (VA) to the customer, or not. A description of each of these terms follows:

Implementation
Plan
ST 0
WT 0
RT% 80
AV% 80
VA No

- Step Time (ST) is the typical time each step takes. Anybody who performs that step regularly will know this number.
- Wait Time (WT) is how long the employee waits before taking that step. It may be zero, or it may be several days because they need to collect information or wait for the right time to begin.
- Roll Through % (RT%) is how often the step is successful the first time. The goal is to get every process step performed correctly the first time. We call the percentage of times the step has to be repeated over time "non-rolled throughput yield."
- Availability % (AV%) is how often all the assets required to perform the step are available to you, compared to how often they should be available to you. This could be people, equipment, or information.
- Value Add (VA) meaning does the step add something the customer would pay for?

Note that we've established a standard unit of measure—days—and indicated it on the map. This map helps to identify obvious areas where improvements can be made throughout the value stream.

Notice that most steps have an upstream supplier and a downstream customer. Upstream suppliers represent anyone executing the process step just before yours. Their work may be to gather data, or something more tangible, and give it to you. We call these "Supplier Processes."

There are several customers in this scenario. There is, of course, the customer at the end of the value stream. There is the employee at the next process step, who is also a customer. A third customer is another employee who may take over the job when you are absent or leave the company. And finally, if you work in a regulated industry the industry auditors are an additional customer. Any of these processes receiving assets or information from your process we call "Receiver Processes."

We'll now take a closer look at interacting with upstream suppliers and downstream customers.

The Upstream Supplier

Upstream suppliers provide the materials and information to you, which you need to complete your work. This makes you their customer. We call this relationship between you and all your suppliers the Supply Chain. Managing the quality of these incoming assets is an important step in a flowing process.

Supply chain interactions go through three evolutionary phases. In the first phase, you must check everything being received to be sure it is correct. Any incomplete or inaccurate deliverable must be returned for rework.

In the second phase, the upstream supplier should be taught about quality, and how to apply quality concepts to deliver improved goods. Eventually, when the quality levels of the incoming deliverables become consistently acceptable, and the completed quality checklists arrive with the deliverables, it is only necessary to spot check the incoming deliverables.

The third phase is the most streamlined. This is where the supplier's quality standards are proven consistent enough that the incoming packages don't need to be quality checked. At this point they just need to have the completed quality checklists arrive with the deliverables, as proof of supplier quality control. It takes time, but this state is what you want your supplier relationship to mature into. If a failure detected in production ends up being attributed to a phase three supplier, you must restart the process at phase one again.

All these activities fall under the Incoming Quality Assessment (IQA) efforts. It's important to nurture the relationships with your suppliers so that you can trust each other and help each other.

Deming often asked companies these types of questions:

- "How have you helped your suppliers better cater to your needs?"
- "Have you shown your suppliers how you use their products?"
- "Have you shared your quality practices with your suppliers?"
- "Have you passed customer recommendations on to your suppliers?"

The Downstream Customer

When interacting with the downstream customer, whether it be the business customer at the end of the line, or simply the next employee receiving your work package, you should always be interested in sampling the customer's satisfaction level.

There are several dimensions to customer satisfaction that have been refined through extensive research in the 1980's and 1990's. The SERVQUAL model provides these five dimensions of customer satisfaction that you may want to include in your employee training and customer surveys. These five dimensions have the greatest impact on the customer experience as they consume a service:

- Reliability: the ability to perform the promised service dependably and accurately.
- Assurance: the knowledge and courtesy of employees and their ability to convey trust and confidence.
- Tangibles: the appearance of physical facilities, equipment, personnel, and communication materials.
- Empathy: showing consumers that you care about them; individualized attention to customers.
- Responsiveness: the willingness to help customers and to provide prompt service.

When building any kind of customer satisfaction questionnaire, these five measures are foundation attributes.

Other factors affect the customer relationship. Propinquity, or how personal you are with your customers, will increase the emotional connection to your brand and company. Attunement, or your listening position, also impacts the relationship. Are you rushed and sharp with your customers, or are you warm and friendly and genuinely interested in how well they are being taken care of?

In addition, two excellent feedback questions to ask customers are:

- "How would you rate your most recent experience with us?"
- "How inclined are you to recommend our services to others?"

These small differences become competitive advantages. Be sure you provide feedback to Engineering and Product Management, as needed.

Chapter 4 - Six Forms of Process Improvement

A simple value stream may have only one repeatable process, but more often a value stream is a chain of repeatable processes. In these scenarios, value flows like a river through multiple processes cascading from supplier processes to receiver processes. Each supplier process produces a deliverable that is received by a receiver process at a checkpoint, added to, and then handed off to a receiver process at the next checkpoint. There is always information flowing along with each deliverable. This information holds the key to process improvement.

Value Realization doesn't happen until the final process finishes at the end of the value stream, and the external customer receives and pays for the final deliverable. This is an important concept. All the effort going into a value stream is either an investment or waste, depending on whether the customer receives the product and pays for it. If time and effort are put into a deliverable, and the customer never receives it, it was waste. If the customer does receive it, that time and effort was an investment.

The key to streamlining a process is to maximize those elements that are of most interest and value to the customer, and at the same time, minimize the time, effort, and cost it takes to deliver those elements. Improving performance over time is the key to lowering costs. Don't spend too much time looking for lower-priced suppliers and lower priced materials. Your big savings will be found in improved processes.

We call the ability to deliver unimpeded value "Flow," and we call the improvement efforts performed by everyone within the company to reduce time and costs involved "Lean." In addition, non-conforming process issues will arise, and we'll need to discover their root causes, triage them as fast as possible, and then mistake-proof them so they don't recur.

After we are sure we are producing the right product for the customer and producing it correctly, we can work on producing it faster and cheaper by using better ideas, and by minimizing the Hidden Factory.

You can't manage all of this correctly until you first systematize the process. How will you know the process is improved if you haven't mapped it out or measured its previous cycle times? How will you preserve a better idea long-term if the improvement is not written down?

How will you save the time spent identifying the root cause for an asset failure going forward?

There are six different approaches to improving a process, all of which contribute to its overall capability. They are listed here:

1. Systematization: Process Modeling and Control
2. Measures: Defining Process Metrics and Service Levels
3. Flow: Identifying and Eliminating Bottlenecks
4. Lean: Identifying and Eliminating Waste
5. Resilience: Addressing Issues as They Happen
6. Durability: Mistake Proofing the Process

All these areas require innovation and continuous improvement, or Kaizen. We will examine each one below.

1. Systematization: Process Modeling and Control

A process must first be modeled, or systematized, before it can be sustainably improved. If an improvement is made on an un-systematized process, the improvement is only an anomaly that will soon be forgotten as soon as a new operator is in place. For this reason, processes must be systematized to be continuously improved. Systematizing a process means to create a proper structure around it, and then to document this structure in a version-controlled repository. The knowledge created in this manner is called "Process Controls."

The challenge in IT is that technology changes so rapidly, driving industry changes, that internal IT processes end up changing more often than most other industries. For this reason, documentation should be minimal. Use diagrams and drawings instead of long pages of text, where possible. When modeling your processes and associated process data, these attributes should be identified:

1. The process name.
2. All triggering events that start the process.
3. The process value propositions, or deliverables of value.
4. Which value-streams the process serves.
5. Which customers the process serves.
6. Which suppliers the process requires.

In addition, each process should be listed in the Process Controls section of the Configuration Management System governing your area, and should contain the following components:

1. A Standard Operating Procedure (SOP) preferably in diagram form.
2. A Standard Work Checklist.
3. A Process Recovery Model.
4. Links to Asset Recovery Models for the problematic assets within the process.
5. Scrap, rework, and cycle-time metrics.
6. A Service Level Goal or Agreement that can be measured.

If you don't have a Configuration Management System, or Configuration Management Database, you'll need to establish one. Think of it as a central repository for storing process and asset knowledge.

Initial Setup

The initial setup happens during Quality Planning, when the initial quality system is being established in an environment. During this initial setup, and periodically afterward, the following should happen:

1. Be sure all repeatable processes that need quality protection have been identified and added to the Process Controls.
2. Be sure all process SOP's are up to date.
3. If necessary, diagram complex flows using flowcharts or sequence diagrams.
4. Add new risk items to process checklists.
5. Improve existing steps so there are fewer mistakes.
6. Innovate better ideas to save work and time.

It is convenient to store this Process Control information in the Configuration Management System (CMS). The CMS should be centralized and easy to access. Quality Planning should be revisited and brought up-to-date every six months or so.

To Systematize an existing process, we'll use the Standardize-Do-Check-Act (SDCA) model. SDCA is based on the more popular Plan-Do-Check-Act (PDCA) model which Walter Shewhart created, and W. Edwards Deming wrote about in his books. While PDCA is a pattern for continuously improving a standardized process, SDCA is a pattern for standardizing a current process that has not yet been modeled.

SDCA was introduced in Masaaki Imai's 1997 book *Gemba Kaizen*, as a tool for standardizing a process. The circular pattern works like this:

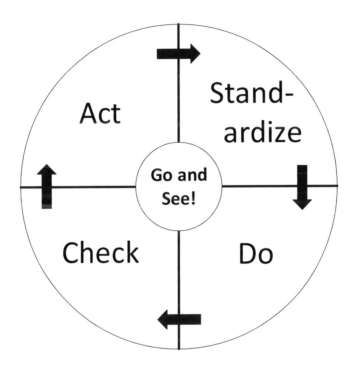

Figure 9 - Standardize-Do-Check-Act

1. Standardize – Create the process model.
2. Do – Execute the process (perform one Process Kata).
3. Check – Check the process model against the process steps for accuracy.
4. Act – If the model is not accurate, improve it.

This pattern repeats until the Operator is confident that the model reflects the repeatable parts of the process accurately.

According to Mike Rother's book *Toyota Kata*, Toyota added "Go and See" to its PDCA and SDCA cycles because the automaker found this to be a critical step. Although it's really part of "Check," Toyota learned it is critical to go to the customer, or to the work location, and see the results in their natural state.

System Learning & Improvement

Once processes have been systematized, they must be kept up to date and improved over time. Defined processes are supposed to represent the best-known way to perform process actions. If a team comes up with a better way to perform a process, the improvement should be adopted, and the process model should be updated to represent the improvement. The whole point of standardizing a process is to preserve institutional knowledge while ensuring consistency and sustainable improvement.

In addition, process reviews should be scheduled every six months or maybe once a year as part of regular Quality Planning to be sure they are accurate.

To improve an existing process, we'll use the Plan-Do-Check-Act model shown below.

Shewhart and Deming used this model to gain what they called "Profound Knowledge" about how their systems worked. A system of Profound Knowledge is a system that is understood by its operators.

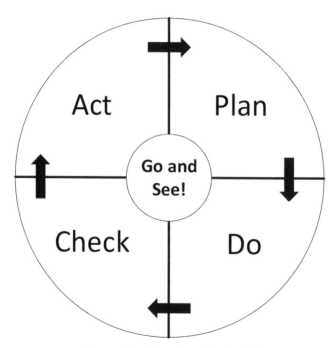

Figure 10 - Plan-Do-Check-Act

For example, Starbucks Coffee owners have learned if they periodically have the sound of coffee grinding in the background, patrons in the store buy more coffee over time. This is a beneficial input to a predictable system they have come to understand. You could say, they have profound or insightful knowledge that this technique works because they have previously experimented with it.

Deming said operators should experiment with different theories until they have converted their theories into profound knowledge of how the system reacts. This empowers them to tune each step for increased efficiency within their processes.

Deming taught us that these four steps are key to understanding a system:

1. Thinking of all the moving parts as a system.
2. Understanding the amount and range of variation occurring within the system.
3. Knowledge of the causes of variation within the system.
4. Understanding the psychology of people working within the system.

He calls his model the SOPK, or System of Profound Knowledge.

The approach is to take a hypothesis like the coffee grinding idea mentioned above (Plan), try it at several locations (Do), and then measure the results (Check). If the results are consistently favorable, start grinding coffee every few minutes in all the store locations (Act).

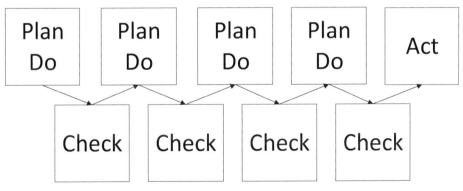

Figure 11 - System Learning Using PDCA

Now, you could adjust the frequency of the coffee grinding over time and find the best tradeoff between grinding frequency and purchasing trends. In a case like this, you would really be performing the PDCA cycle multiple times and finally selecting the optimized results. This is called 'System Learning" and the pattern will look like the previous diagram.

Non-conforming Output

If a defect escapes and is found by a customer, the organization is faced with three possibilities. A specific new external problem may have caused the defect to occur, and after a root cause analysis is conducted to determine the reason, the process can be protected against the cause by removing the vulnerability completely, or by adding a "watch out for" item to the process checklist. We call these "Special Causes."

The second possibility is that natural variation within the process created an unexpected result. These types of problems usually require solution rework from the engineering group. We call these "Common Causes."

The third possibility is that the operator did not fully understand an item on the checklist and checked the item off without completing it. In this case, the operator will require retraining with the process controls. This retraining is an opportunity to re-read the Standard Operating Procedure and to update it and the process checklist for clarity, if necessary. This re-training should be recorded in a historical training log.

2. Measures: Defining Process Metrics & Service Levels

If it's not measured, you can't improve it. At least, not sustainably. Not everything needs to be measured, but for processes in need of improvement, the starting point is measurement. Measurement is a doorway into reality and dispels bias. Measurement is a great way to make a persuasive argument and simplify disagreement.

Metrics should be a visible part of the organization's culture and should be used to trigger brainstorming sessions with positive dialogue for better ways to improve existing practices.

Historically in Western culture, measurement has been a way of conditionally evaluating employees. This has produced much anxiety around metrics in the workplace. Care must be taken to reinforce the message to the group that the purpose of measurement in a quality system is to connect us to reality. With accurate measurements, we can work together to brainstorm ways to get more work done faster with less effort and resources.

Process Metrics

One simple rule for measurement is that the results should be similar regardless of who performs the measuring.

There are several natural attributes that can be measured for almost any process. They are:

1. Customer expectations.
2. Customer satisfaction with the process results.
3. Rate of customer consumption.
4. End-to-end process cycle times.
5. End-to-end process costs.
6. Amount of scrap and rework over time.
7. Post-sale warrantee losses over time.
8. End-to-end process capability: Did it work every time?
9. Output variation over time, using a Statistical Process Control Chart.
10. Root causes of failures resulting in scrap or rework, using a Cause and Effect Diagram.
11. Frequency and priority of failure causes resulting in scrap or rework, using a Pareto diagram.
12. History of the triggered process, customer, or market events, using a Histogram.
13. External Impacts on process outcomes, using a Scatter Plot Diagram.
14. Process challenges and their sources, using an Interrelationship Diagram.

In some situations, measurements can be made more useful by combining two or more measurements. For example, measuring customer adoption rates can be insightful, but measuring customer adoption surges after specific marketing campaigns can be much more insightful.

Reporting the number of bugs found is not as helpful as combining that number with the percentage of modules having accompanying test cases, and the percentage of test cases completed.

The various types of charts most helpful for each of these measures are described in the last chapter of this book titled "Data Collection and Analysis Tools."

This is only a sampling of common process attributes that can be measured. Care must be taken to identify a business need for the measurement, and only spend time measuring process attributes that require optimization. In other words, don't spend time measuring something that has no business value.

Service Levels

A Service Level is a desired measurable level of performance achieved by a process. It may be as simple as indicating if a process has completed successfully, or it may be something complex like measuring the uptime of a service during defined business hours per year.

If there is no formal contract with a customer, we call this target a Service Level Goal (SLG). If a formal contract with the customer exists specifying a level of performance, we call it a Service Level Agreement (SLA). In addition, an organization may create an internal Service Level Agreement for a service offered with a publicly promised standard.

3. Flow: Identifying and Removing Bottlenecks

Flow is the lack of resistance to movement throughout the value stream. Good process flow enables better predictability, less batching of work, and continuous delivery to the customer.

Flow is important within a value stream so that customer orders do not get delayed. If customers are delayed too long, they will go elsewhere for business.

Often the workflow must be completely restructured to increase the flow of information and assets through an organization. Modern manufacturing facilities are designed from the shipping bay first and back to the materials input bay. People that were originally organized into groups of roles become re-organized into cross-functional teams containing each role necessary for the work to flow smoothly. This restructuring makes it easier for flow to move through the organization and is a huge competitive advantage.

Think of the old pin-ball games. You want value flowing through your system directly from supplier to customer, instead of hitting all the bumper stops and bouncing back and forth, from silo to silo. This direct flow requires a new focus on cross-functional teams, instead of traditional, functionally siloed departments.

Recently, I was talking with a group that was asking for better tools for Operations, and DevOps. I told them they needed better team structures, not just better tools.

One-piece Flow

Also called "Continuous-flow," or "Single-piece flow," One-piece flow is a work configuration accommodating the complete construction of one sellable item from start to finish with as little work-in-process inventory accumulating between operation steps as possible.

Contrast this with batching multiple production items during construction. In a One-piece configuration, each single item is ready for sale much quicker than the entire batch will be. In contrast, items for sale in the batch must wait for every other item in the batch to be completed before any of the items can be sold. Waiting ties up resources, capital and delays sales income. In IT this is true for Portfolio Management, Development, DevOps, and Implementation.

The Theory of Constraints

In 1992 Dr. Eliyahu Goldratt wrote a ground-breaking book called *The Goal.* In his book, he tells the story of a plant manager frantically trying to make a failing plant become profitable again. He tries a lot of conventional ideas such as maximizing resource utilization and running double shifts but quickly finds out these techniques cause more problems than they solve.

He finally discovers the real "goal" is to examine the entire value stream and identify the slowest bottleneck in it and then fix it, continually repeating this cycle until work can be completed at the same rate of customer consumption.

He presents the idea that the cost of the bottleneck is the cost of the entire enterprise idly waiting for each work item to move through it. He calls this the "Theory of Constraints" (TOC) and provides a 5-step plan for improving workflow within a value stream.

Five Steps for dealing with Constraints:

1. <u>Identify the biggest constraint</u>: Think of this as strengthening the weakest link in the chain. A focus on other links is not as beneficial while this weaker link exists in its weakened state.

2. <u>Exploit the constraint</u>: This means to do everything possible to minimize the effects of the constraint, preferably using available resources and taking care to not spend more money in the beginning to solve the problem. Be sure the step containing the constraint has priority with work items and attention so that it is always executing, and its slowing effects are minimized. Also, be sure that as much quality control as possible happens before the constraint so that known defective items do not get sent through the expensive constrained step, only to be reworked later.

3. <u>Subordinate everything else to the constraint</u>: Line up all dependent activities and have them timed in production with the constraint, and no faster. Producing more upstream incoming work items than the constraint can handle creates waste because these intermediate components will have to wait their turn through the constraint.

4. <u>Elevate the constraint</u>: Once the capacity of the movement through the constraint has been maximized using the steps above, the next logical step is to invest additional resources into expanding the constraint. It may require more people, or tools, or money. Be careful not to jump from step 1 to this step, skipping steps 2 and 3. A lot of lift can come from steps 2 and 3 before needing to spend additional money on the problem.

5. <u>Repeat steps 1-4</u>: Repeat these steps until your rate of production matches the rate at which your customers consume your products or services.

I remember one client explaining to me how his or her team of 50 programmers would wait as eight QA testers spent five days regression testing the entire system, every three weeks. Supposedly, the developers were busy with new work during each third week. In reality, their production was somewhere around one third of their true capability during this time, as they waited anxiously for the inevitable bug reports to be given back to them.

This was a classic TOC problem. Regression testing was the biggest constraint (step 1). One solution was to have the developers peer test each other's work before it was sent to be QA Tested (step 3). This would eliminate a lot of the rework coming back from the testers. The second part of the solution was to elevate the constraint (step 4) and have all 50 developers become testers for 1 day (steps 2 & 4), reducing excess input (step 3), and shortening the regression testing process from five days to one.

4. Lean: Identifying and Removing Waste

Lean Manufacturing principles have been around for about half a century and are still being adopted aggressively by manufacturing facilities. Lean, in a sentence, is identifying and eliminating waste within a value stream. Every activity within a value stream either adds value to the finished product or it doesn't. If it doesn't then it contributes to the Hidden Factory of waste, called "Muda" in Japanese. There are eight types of activities that consume resources of some type but do not add value to the final product. Together they spell the acronym T.I.M.W.O.O.D.S.

Eight contributors to the Hidden Factory of waste (Muda):

1. <u>Transport</u>: Transport is the time delay waiting for an asset as it is moved from one location to another. In modern warehouses, this is easily identifiable by the elaborate conveyers spread out in some facilities, and the carting needed to move items around in them. In IT this can be physical transport encumbrances such as shipping a server to a customer site, but more often transport waste occurs when unnecessary knowledge transfer is needed. Employee attrition, off-shore development, filling someone in on details

because they were not part of a discussion, or transporting requirements onto paper to be read by a design team that was not at the discussion are all examples of transport waste in IT.

2. Inventory: Inventory costs money. It costs to rent the space to store excess material. Fires, theft, and revisions can destroy inventory. There is a natural law that says higher inventory levels conceal problems. If there is a large supply of needed parts, nobody will spend the time identifying just how many are required to meet actual demand. In this case, efficiency is lost. In IT, a heavy stack of requirements is inventory waste. They will rot away into irrelevance over time, as the customer's needs change.

3. Motion: Any human motion not directly related to adding value is unproductive. A person walking or reaching beyond his or her workspace to obtain a tool is creating waste. For this reason, the more frequently a tool is used, the closer it should be to the workers reach. In a knowledge-working environment such as IT, think in terms of how easy information can be seen. What type of screen layouts do the team members have? I used to think writing code with dual monitors was a treat until I tried using eight monitors simultaneously. That changed my life.

4. Waiting: Waiting happens when the hands of the operator are idle such as in meetings, while waiting for decisions or pending approvals, or when needed information is uncollected or ambiguous. Waiting on requirement elicitation or clarification is a form of waste.

5. Overproduction: Overproduction means doing more work than is necessary right now. In IT, this is often called "gold plating" and can be described as adding features that no paying customer requested, or doing additional activities that are not required. This can also be making decisions without the proper representation of stakeholders present, which often ends up causing rework or scrapped work when the other decision participants become available and have different ideas.

6. Over Processing: Waste in processing can happen when processes take an unnecessarily long time to complete, or if two processes could happen in parallel, but for some reason don't. Task switching is time wasted moving between two activities and is a form of over processing.

7. <u>Defects</u>: Defects interrupt production and require rework or replacement.

8. <u>Skills</u>: Under-utilizing the capabilities of your people, lack of teamwork, or inadequate training all represent skill waste.

5. Resilience: Addressing Issues as They Happen

In addition to Flow, and Lean techniques, Process Issue Management is a critical part of building a robust process.

When problems occur, and they will, the operator needs to get the process back on track as soon as possible. In order to do this, we use Process Recovery Models, and Asset Recovery Models, as part of the process controls.

A Process Recovery Model is a list of known problems that have occurred in the past with the process. It contains a description of the problem, a root cause, and a solution to recover from the problem. This enables operators to quickly be able to determine if a problem that was just triggered has a known solution. This saves the time needed to diagnose and come up with a solution when someone in the past has already spent time doing that. Process Recovery Models should be easy to access and readily available.

Asset Recovery Models are the same as process recovery models except they are based on an asset within the process, which may be used by other processes.

An example would be an application server used by a call center. Occasionally a clustered server may failover to the secondary node, alerting the operators that something is wrong.

In a situation like this, the operator would be trained to go directly to the Configuration Management System, obtain the Asset Recovery Model for the server, look for this particular problem, and see what root causes were found in the past to explain the server's state. Remediation steps should also be listed, to enable the operator to quickly fix the situation and get the server back on its primary node.

When a new problem is encountered, a root cause analysis would be conducted, and when the source of the new problem is identified its description and remediation steps would be added to its Process or Asset Recovery Model.

Root Cause Analysis techniques are covered in the Process Owner chapter.

6. Durability: Mistake-Proofing a Process

W. Edwards Deming said that management should stop blaming employees for process issues. Instead, the burden of mistake-proofing a process should fall on the management team. The management team should work with the operators to brainstorm ways to prevent the same mistakes from occurring in the future.

Elimination

Eliminating unnecessary or error-prone steps is a great way to reduce errors. Work with your supplier or select a different sequence that could result in a simpler set of required steps to produce the same results.

Simplification

The reduction of variation is always a good idea. Variation, by itself, creates overhead of some kind or another. Sometimes that overhead becomes an exponential resource drain.

Many companies have discovered if they can reduce internal variation, they can become more competitive. In IT, a simplified set of database platforms, programming languages, and architecture components means fewer skills are needed for development and support activities, for example.

Automation

In IT, we often automate tasks to remove manual redundancy or protect the process against human error. Many sophisticated and inexpensive automation tools exist today. Automation, where practical, is always a good idea.

Acceptance Screens

An Acceptance Screen is a mistake-proofing tool that prohibits an asset from proceeding through a system and onto the next step until it is structured correctly. Toyota called this "Poka-Yoke" which is a Japanese term that means "Mistake Proofing." It was made popular by Shigeo Shingo and is a helper-mechanism made to reduce or eliminate possible mistakes.

The plastic shim on the inside half of a USB drive, for example, prevents a user from inserting the drive into the USB slot incorrectly. Many ATM machines require you to take your card back before they dispense cash.

Bank software applications use a check-digit technique to dramatically lower the probability of errors when typing in long account numbers. A check-digit is a number added to the end of an account number which is calculated from a simple math formula applied to the account number. Thus, the receiving computer having the same standardized formula can perform a check-digit calculation and compare the result to the digit at the end of the account number to determine if the entire number is a valid entry. All your credit cards have standardized check-digit formulas that are published.

Acceptance Screens can reduce the Hidden Factory without requiring much cost to implement. Shigeo Shingo, while working at Mitsubishi discovered that using Acceptance Screens where possible significantly reduced the need for statistical process control, which reduced the need for measurement and analysis.

Chapter 5 - The Components of a Quality Program

My friend John Lee wrote a book called *Rising Above it All* which won the Shingo Research and Professional Publication Award in 2016. In his book, John talks about two different approaches to achieving a high-quality process.

The first and most common approach is to hire competent people and then pressure them. At best you might achieve something like a 95% process capability. Process Capability means how many times a process is executed perfectly, with no rework, scrap, unwanted variation, or unexpected delays. A Process Capability of 100% means every step of the implementation is successful the first time through. This requires people with competence to get it right as much as humanly possible, and tough skin to take the scolding when things don't go right. John calls this PBP, or Personify, Blame, then Punish. You will get pretty good results with this technique as you churn through good employees, but you will never be world class.

The second and more effective method is to create a systematized process to govern the work performed, and then have everyone use the process, and keep improving the process until your capability is more like "five 9s."   Five 9's means your system is working without problems 99.999% of the time. This is how companies achieve world-class process capability, and world-class service provider status.

A systematized quality program is like a symbiotic helper. The employees and management work together to continually improve the process, and the process reciprocates by improving the performance of the employees and management.

What is a Systematized Quality Program?

It may help to first define quality. As a word, quality means the degree of excellence of something. Of course, the mark of excellence is open to interpretation. Deming taught us that quality must be defined by your customer. Therefore, a simple way to think about quality is to produce something that meets the needs of your customer exactly. Remember that you often have multiple customers, and that their needs constantly evolve. Part of the challenge of defining quality is being sure that your understanding of your customers' needs is kept current.

A quality program is a collection of structures created to enable imperfect people to sustainably produce quality products and services. A quality program manages input quality, throughput quality, and output quality. We call these three areas Incoming Quality Assessment (IQA), Quality Assurance (QA), and Quality Control (QC), respectively.

There are two other components called Quality Planning, and Continuous Improvement. Quality Planning is where we set up the whole program, and Continuous Improvement is a set of mechanisms we put in place to improve the products, services, and our ability to deliver them over time. We'll discuss Quality Planning first.

Figure 12 - Quality Management

Quality Planning

Quality Planning is the systemization of a work environment. It starts with identifying the value propositions offered by a system, and then identifying all the services that makeup the value stream required to deliver those value propositions. It includes the initial identification of the Master Services offered by the environment, and the processes, triggers, inputs, outputs, customers, and suppliers needed to facilitate the services. In addition, the specific metrics that represent goals important to the organization and its customers are identified. All this information gets stored in a Configuration Management System (CMS).

If you don't have A CMS, you'll want to start by creating a central repository for storing all your process information. A good Confirmation Management System is version controlled and easy to navigate.

Quality Planning should be refreshed about every six months.

Incoming Quality Assessment

An Incoming Quality Assessment, or IQA, is a much-overlooked component in Quality Management and is the first step in minimizing the Hidden Factory. An operator should examine incoming deliverables or information from the supplier and be sure it is complete and everything is available.

If something is missing, the supplier should be notified immediately. It may be that the operator can proceed with incomplete material or information, but both the supplier and the operator need to know that each understands there is an outstanding deficiency and the supplier still needs to provide additional inputs.

In some cases, the input should be rejected, or quarantined until the supplier delivers the outstanding material or information. With ongoing supplier relationships, Deming taught us to develop the relationship so that they are consistently giving you quality material or information that meets your system input standards.

Quality Assurance (Process Quality)

Quality Assurance is the proof that the process in question was followed correctly. It is the operators completing checklists during or at the end of their work and then passing them along with the completed deliverables or processes at strategic checkpoints. These checklists become historical logs. Data from these checklists is not used to punish operators for what they did wrong, but rather to provide a glimpse of reality to drive retrospective conversations and brainstorming sessions for how to improve performance going forward.

In time, these processes become improved, and institutional knowledge is preserved. The enabled organization can make better decisions over time, resulting in fewer mistakes, less rework, and less scrap. This enables the organization to produce the same results with less effort. Having these elements in place ensures the organization is a learning organization. This systemization of process steps throughout the value stream provides a foundation for real improvement.

Quality Control (Product Quality)

Quality control is also known as inspection. It is a way to ensure high product quality. When a candidate item is complete, and before it goes to a customer, it should be inspected to be sure it is ready. Quality control has two steps: verification and validation.

Verification means a completed deliverable is checked to confirm it meets the quality goals it was designed to meet, which would be found in the product or service requirements. If for some reason a completed deliverable fails this test we must rework or scrap it. We pay for this fix and more time and money is lost to the Hidden Factory.

Testing software is verification. When testers find a problem, the engineers are notified and must rework the code. In some instances, the entire feature, module, or even application may need to be scrapped and a new approach taken until verification is completed.

Once an item passes verification, meaning we are confident it has met all confirmed requirements, we show it to the customer for inspection and solicit final approval. This is called validation. If the customer accepts the item, we are finished creating it. If the customer wants something changed, a change request or change order is required to alter the item and the customer must pay for the rework.

In software we call this validation process "User Acceptance Testing," or UAT.

Does it match our requirements?	Verification
Validation	Can the customer use it as intended?

The same quality control principles apply to any deliverable, regardless of whether the customer is the final business customer or the customer at the next step in the process.

Continuous Improvement

The first goal of any organization is to deliver exactly what the customer wants. Measuring customer satisfaction levels and observing customers using a product or service provides immediate feedback on how well the organization is meeting its customers' needs. The problem is that customers' needs change over time.

Market trends evolve, technology improves, competitors emerge, and many other dynamics impact a customer's reality over time. For this reason, relentless surveying and observation is required to be sure the company is continually servicing its customers accurately.

Some improvements can be made by the suppliers to a value stream. Any suggested improvements to process inputs should be communicated upstream to the process suppliers. This requires a healthy dialog between a process operator its process suppliers.

In addition, while companies are delivering value to their customers, they must also be reducing the Hidden Factory. Organizations with the smallest Hidden Factory are naturally the most competitive.

Learning mechanism are needed to facilitate these continuous improvement trends within an organization. Customer satisfaction surveys, observation, supplier dialogs, measurements, retrospectives, root-cause analysis, experimentation, failure, and learning from the past are all part of Continuous Improvement, and all these activities produce knowledge that must be fed back into our process controls, making the systems more efficient and effective over time.

The only way an organization can be a real learning organization is by preserving institutional knowledge. The information inside people's minds that is not written down is tribal knowledge, and as soon as they leave the department or organization that knowledge leaves with them. Therefore, important information, lessons learned, and process improvements should be stored and updated in the Configuration Management System (CMS).

The CMS must be stored in a centrally accessible and known location so that it can be easily accessed. Many companies augment their CMS with an internal WIKI.

Training

Quality programs require training on how the program works, and training on the complex assets the operators will be using while working within each process.

Training should be logged, and periodic retraining may be required every six months or perhaps annually. Employees new to the program should be trained, and anybody caught not following the quality system should repeat the training.

Non-Conforming Product Management

Another component in a formal quality program is a list of customer-reported issues with a product or service. In IT, this list is typically created by the Service Desk. The Service Desk should be separating individual issues from repeating issues and bringing the repeating issues to the attention of the Engineering or Operations teams, as needed.

The Service Desk, if equipped properly, can become a central portal for existing customers to interact with your organization. A good Service Desk can make a poor product passable. A bad Service Desk can make a great product suffer.

While the Service Desk can take care of minor issues and configuration problems, more serious issues should be forwarded to the CCAPA Committee for triage. CCAPA is an acronym for Correction, Corrective Action, and Preventative Action. The CCAPA Committee decides how the identified problems are prioritized for rework and should be staffed by a representative from Engineering, Operations, Implementation, and the Service Desk.

The CCAPA Committee should meet regularly, and in some cases such as just after a major release, they should meet daily.

Measuring the Cost Savings of Quality

There is, of course, an overhead cost incurred in time and money for performing all these quality activities. It takes time to create Standard Operating Procedures and the checklists, and then review them regularly. It takes time to meet with other busy employees, make appointments and meet with customers.

The alternative is an environment filled with unscheduled problems causing rework, scrap, fires to put out, angry customers, reputational damage, and lost revenue.

Quality Management is not about adding unneeded overhead to your daily work, it's about trading predictable, pre-scheduled time for not having the downstream unpredictable, un-scheduled time burdens needed to recover from the never-ending circus of reactions.

Teach your people to respect and improve the system. Build a culture that allows for mistakes, reflection, innovation, new ideas, and group creativity for solving problems.

> Quality Management is not about adding over head to your daily work, it's about trading **predictable pre-scheduled time** for downstream **unpredictable un-scheduled time** needed to recover from the never-ending circus of mistakes.

Celebrate victories together. Generate better ideas and performance improvements as a group. Challenge your team to improve continuously. When mistakes are made, don't blame the employee, instead challenge the individual or the team to come up with a way to prevent anyone in the future from repeating that same mistake.

Won't Heavy Process Slow Us Down?

When W. Edwards Deming first approached American car manufacturers in the 1940's and told them he had techniques that could double the lifetime of their products, they dismissed him. They had considered other early industry attempts to manage quality by past failed companies, and rejected the idea that it should be a prominent focus.

In 1949, Deming was asked to help with the joint U.S-Japanese census efforts in Japan after WW II. While there he was able to share his ideas with Japanese manufacturing companies. He stayed in Japan for 25 years helping Sony, Toshiba, Toyota and many other companies rebuild their factories around his ideas. Together, they created a collection of guiding quality management principles we now call Lean.

In the end, they proved the American car manufacturers wrong. They demonstrated they could produce a higher quality product in less time and for less money, and they did it without expensive robotics and technology. The difference was that they had learned the correct way to manage quality.

The same argument surfaces today. Often, technical teams approached with a quality management proposition push back, thinking the process will slow everything down to a crawl.

It is critical to understand that a quality program presents a trade-off in time and resources. Some overhead in time is required to operate within a quality program—minimal if executed correctly—but the benefit of the trade-off is that your team will have much more time freed up going forward due to the minimization of unexpected rework, scrap, delays, and downtime.

You will get some short-lived pushback from newcomers when establishing a systematized quality program, but once they have a few positive experiences from their quality monitoring efforts they'll catch on and become believers.

Good Versus Bad Process

A good process implementation does not provide much resistance to the natural flow of work through a system. Even though checklists are used, and surrendered at checkpoints, the workflow should never be held up waiting for an approval. The key is to make sure the same people making the decisions are the people with access to the information.

By contrast, bad process has a noticeable distance between the decision maker and the information. This is a common source of artificial delays.

One example of bad process occurred during Desert Storm. Some colleagues of mine worked in Iraq as government contractors during the war. They explained that for some approvals they had to suit up, get into hardened vehicles, drive from their base through several known kill zones, and into another base where the higher-ranking officer was available, only to obtain a signed piece of paper authorizing something. Then, the team had to drive back through the kill zones to be able to proceed with the next step in their process. Although this process needed improvement, these men were clearly dedicated.

A Quality Program is Optional

The US Auto Manufacturers were faced with this decision in the 1940s, and said no. As a result, the Japanese who embraced these practices put the US Automakers into bankruptcy, requiring a bailout from the US taxpayers. They are going bankrupt again, just much more slowly.

A quality program is optional, but not if you want to stay competitive.

My First Experience with a Quality Program

One of the companies I worked for as a software development manager had over 200 sites using our client-server software. It was a complex system with a lot of moving parts, and consequently, we had a lot of unexpected problems continuously appearing.

I remember after some time, we started budgeting future schedule bandwidth for unforeseen problems that we didn't know about yet but were sure we'd have in the future, because we seemed always to have had them in the past. Then, when the next week rolled around, we would congratulate ourselves because we were right—a whole new slew of issues was now on the table to address. These were usually not coding issues, they were implementation problems.

One customer had an old version of Adobe Acrobat and that was messing up something else. Another customer, somehow, was missing several tables in the database and we had no idea how our system was even working at that customer's location. With more and more customers, there was an endless river of problems we had to deal with. Finally, our wise CEO said, "I've seen these kinds of problems before—we need a quality program."

That same day he called some consultants who came and spent a year with our company. They went from department to department helping us understand what a quality program is and how to set one up. We identified all our repeatable processes, created Standard Operating Procedures, checklists, and checkpoints, and nothing could advance until it met all the criteria at each checkpoint.

We complained about the extra overhead and discipline required for this level of accountability. Some of our best people even fudged the checklists in the beginning. Our production was slightly slower, at first.

But then, something interesting happened. Our quality system started catching potential failures before our work mistakes ever made it to the customer. We started finding problems in things we would have otherwise skipped over as we rushed on to the next process step, or into production. Suddenly, we realized everyone needed to be upgraded to the latest version of Adobe. Our automated code-build process needed to have a few extra steps added to it. Databases needed to be checked for all tables every time they were upgraded. One by one our problems disappeared and didn't return.

We started celebrating our victories. We made a list and announced to our entire department every time our quality program protected us from a future problem. We posted this list publicly and had conversations speculating on how much additional time each potential mistake would have cost us.

A short time later, after the effects of our Quality Program began to permeate down to our customer base, we were in a new reality where everything was calm, and we had no fires burning. I remember distinctly asking the others in the department during a meeting if they could think of any problems we were dealing with outside of our current project work efforts. Nobody could name a single problem. We now had much more time to devote to our commissioned project activities because we had no fires to put out. We were able to complete more project work each day going forward.

We had transformed our environment from a circus of reactions into a stable, progressive, learning organization.

Chapter 6 - Common Quality Programs

Several popular quality programs exist today. They have been around for some time. They have a fair amount of overlap yet contain some unique qualities. Here is a brief summary of each:

ISO 9000x

The International Organization for Standardization published the ISO 9000x series of quality standards for manufacturing in 1987. Since then the organization has grown and produced many variations of its standard tailored for various specific industries. At its core, these quality standards are comprised of Standard Operating Procedures, Checklists, Checklist Historical Logs, Non-Conforming Item Registries, and Formalized Training, all of which are tracked for all participating employees.

Lean

Lean is a collection of practices that prescribe identifying and eliminating waste within a system and keeping a workplace clean and organized. Taiichi Ohno, who was the chief developer of the Toyota Production System, is credited for the origin of many lean practices.

An MIT Student named John F. Krafcik coined the term "Lean" in a 1988 thesis paper titled, "Triumph of the Lean Production System," which he wrote after a visit to several Japanese manufacturing plants.

In his opening paragraph of his paper he wrote, "Assembly line workers perform not only production line tasks, but also quality control and preventive maintenance activities."

In 2004, John Krafcik became the CEO of Hyundai. While there, he helped the company increase its quality levels to eventually win the 2015 best car brand for quality, according to J.D. Power ratings.

Six-Sigma

Six-Sigma is a set of tools based on Statistical Process Control targeted towards identifying ways to locate, measure, reduce, and eliminate variation within a business. Companies have learned they can save money by identifying points of variation within their value streams and then reduce that variation.

For example, if you can reduce the variation in lead time from a supplier, you can stock less inventory needed to fulfill the same number of orders over time, freeing up valuable warehouse space, and pre-sales capital.

In software, if you track production errors found per module lines of code, you'll discover an 'ideal lines of code' size where production errors tend to be minimized. Going forward, your teams could create a guideline limiting module sizes to that recommended size, minimizing post build errors across the board.

Six Sigma was introduced by Motorola in 1986 and is often coupled with Lean practices. Today it has no central governing body.

ITIL

The Information Technology Infrastructure Library (ITIL) was originally developed by the UK Central Computer and Telecommunications Agency (CCTA), in the late 1980s. It is a set of practices designed for managing assets and the Service Desk in a typical IT environment. It contains reporting tools such as a Service Catalog—showing the status of all services—and a SLAM Chart showing the performance of a service against its agreed service level over time. ITIL also introduced the Asset Problem Recovery Model, a quick-check list of lessons learned from past failures with a particular asset that can be used by operators to quickly troubleshoot a production asset, minimizing downtime.

ITIL emphasizes the practical value of the Service Desk in an IT organization. The Service Desk is described as a central interaction point between customers and the organization and can be tapped to provide product improvement insight and customer satisfaction information.

CMMi

CMMi is the Capability Maturity Model integrated, a product of the tax-payer funded Software Engineering Institute (SEI) at Carnegie-Mellon University in Pennsylvania. CMMi is a 5 Level model of Software Development Process Maturity containing checkpoints for approval and regular retrospectives. The origins of CMMi date back to 1987.

CMMi promotes five levels of process maturity, and companies can work towards achieving them and then becoming certified as a level 5 CMMi software organization.

COBIT

COBIT is a set of generic processes for managing IT. The "Control OBjectives for Information and related Technologies" is a pre-packaged set of recommended best practices for aligning business with IT. Each identified process contains inputs, outputs, and performance measures. COBIT was first released in 1996 by an international professional association focused on IT governance called the Information Systems Audit and Control Association (ISACA).

Toyota Production System (TPS)

The Toyota Production System (TPS) is a set of practices pioneered at Toyota, who in the 80s and 90s impressed the world with its superior automobiles. TPS is really a set of philosophies about building quality into the production process, instead of only inspecting for issues at the end of the line. This is the basic ingredient to Toyota's success. In the beginning workers check each component against success criteria before it gets added to the assembly line. After they have removed enough variation and the system becomes predictable, they only check one part every so often because it becomes a form of waste to check every single widget of a production line that has consistent precision.

Toyota also pioneered the Just-In-Time practice of requiring suppliers to deliver supplies frequently, just moments before they were needed on the assembly line. In this manner materials were moved directly from the trucks to the assembly lines, eliminating the need to temporarily store them in a warehouse. The primary concepts in the Toyota Production System are we call "Lean" in the West today.

Hyundai/Kai Quality Model

According to a 2015 JD Power consumer report, Hyundai is now the quality manufacture of choice in the auto industry. Outperforming Toyota, Hyundai cars have the fewest reported defects over time, and at the time of this writing, are the best cars to own in terms of quality ratings.

Hyundai used the "Sandcone Model" to out-perform Toyota. Using all of Toyota's very public practices, Hyundai added this one more practice. The Sandcone model is a portfolio management prioritizing guide.

Portfolio Management is the advanced practice some companies use to prioritize which projects are to be funded next.

The Sandcone model states that when deciding which type of project to spend money on, a company should prioritize spending in this order:

1. Customer feedback priorities
2. Product stability and robustness initiatives
3. Decreased production response times initiatives
4. Cost reduction initiatives

Different companies disagree over which order to prioritize those four interests, and there are many debates in academia over which is the best order. After trying several variations, Hyundai chose that one...always listening to its customers first, then stability, then decreased response times, and then cost cutting initiatives. Note that Hyundai was the first car manufacturer to provide USB ports for charging and playing MP3's. Many other car manufacturers still have not caught on to this.

On a personal note, I was at an auto show in Salt Lake City some years ago and told a Hyundai representative her company had great cars, except the company should take a lesson from the Germans and put the emergency brake in the center console rather than the floor mounted push-button design, which Hyundai had at the time. German car manufacturers do this deliberately because it enables the driver to slow down quickly without activating the tail lights, should a police car suddenly appear. This tool is part of the ultimate driving experience. Voila! Today, Hyundai has relocated the emergency brake to the center console. I like to think the company listened to me.

The Stable Framework™

W. Edward Deming, in his book *Out of the Crisis*, wrote about a stable framework that companies should create to systematize and continually improve their repeatable processes over time. Such a framework would reduce ambiguity, errors, rework, and scrap, and enable an organization to produce more value with less investment over time.

Inspired by his term, this book explains how to implement the Stable Framework™, which is a light-weight, out-of-the-box quality program designed for easy implementation, while leveraging the best parts of the above programs. Follow the steps in this book to enable the smooth flow of value within your organization.

Chapter 7 - The Stable Framework™

The Stable Framework™ is designed to handle the operational side of an IT environment. While Agile practices exist to facilitate Development, Stable was created to facilitate Operations. In addition, Stable is applicable to Implementation, DevOps, and any repeatable processes performed within Development. Stable can be used alongside an Agile development team, or it can be implemented as a complete replacement for Agile.

The Stable Framework™ is comprised of one Master Cycle, two roles, three domains, four core meetings, and five values. Let's examine each.

The Master Cycle

The Master Cycle is how we measure progress over time. Like an Agile Sprint, a Master Cycle can span in length from one to four weeks. The length should be consistent, and if you have an Agile team it's best to synchronize your Master Cycle with your Agile team's sprints. This way you can report together at the end of each cycle and sprint.

At the beginning of the cycle, a Cycle Planning Meeting is held that can last up to several hours. In this meeting, the team establishes work to be done during the cycle.

During the cycle, each member performs planned work they have selected at the beginning of the cycle, and any urgent work that has become important during the cycle.

At the end of the cycle, a Cycle Review Meeting, and a Cycle Retrospective Meeting are held. Combined, both meetings at the end of the cycle should last no longer than two hours.

Work activities are prioritized at the beginning of each cycle from several queues: Scheduled Activities, Customer Requests, Asset Maintenance, and the CCAPA Committee. Work is managed using a Kanban system, which allows the participants to stay focused on the highest priority tasks and avoid unnecessary task switching. Repeatable process steps are called Katas and have Process Controls created for them. Process Controls included Standard Operating Procedures, and corresponding

Checklists, called Kata Cards. Performance against process service levels is reported using Service Level Attainment Monitor Charts, or SLAM Charts.

Every day within the cycle the group performs a daily Kaizen Stand-up Meeting, where group members inform each other about the progress of their work activities, and about any improvements they've made in their environment.

Kaizen started as a Japanese word that means "Ongoing improvement involving everybody, without spending much money." We call it a Kaizen Meeting because as we perform our activities, we are always seeking ways to improve them. In 1993, Kaizen was added to the Oxford English dictionary and became an official English word.

The Master Cycle

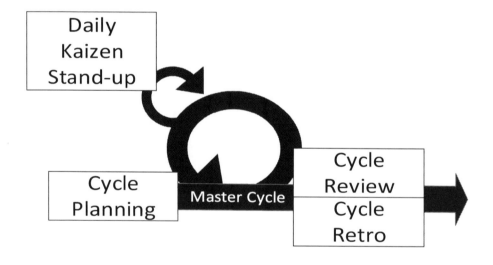

Figure 13 - The Stable Framework™ Master Cycle

It's important for the team to communicate during each cycle. Communication happens in two ways. Horizontal communication happens in the recurring meetings as mentioned previously. Vertical communication happens during normal business activities. Some information is exchanged using horizontal communication, and some using vertical communication.

> **Horizontal Communication** happens in pre-set, recurring meetings.
> **Vertical Communication** happens during normal business activities.

The Two Roles

The two roles in Stable include the Master Chief and one or more Process Owners. The group leader is the Master Chief, and the group members comprise one or more Process Owners.

Project Managers and Agile Product Owners can participate in Stable. They retain their titles, but wear the hat of a Master Chief or a Process Owner. In addition, the team self-selects a facilitator to assist with calling team meetings and representing the team when needed. This role should be changed every so often, such as at the beginning of each Master Cycle.

Master Chief (MC)   Process Owner (PRO)

Figure 14 - The Two Roles in The Stable Framework™

The Master Chief (MC) is responsible for making sure the Stable Framework is set up and followed correctly. Master Chiefs work "on" the system. This means:

Environment Support:

1. Establishing the Stable environment (Quality Planning), including Configuration Management, the three domains, the performance console, and training for all involved.

2. Providing to the team a culture of empowerment, accountability, and continuous improvement.

3. Coordinating the whole set of value streams and their associated processes that collectively we call a System. This is done through facilitating the System Schedule and the Work Queues in the System Backlog.

Individual & Team Support:

4. Enabling the individual success of each Process Owner. This includes establishing the intent, boundaries, and constraints for each process, empowering the Process Owners, and coaching each Process Owner to continuously improve the performance of his or her repeatable processes, called Katas.

5. Prioritizing the work objectives for the team of Process Owners in the Cycle Planning Meeting, and at the start of each day in the morning Kaizen Stand-up Meeting, as needed.

6. Presenting suggestions from the suggestion box to the group each morning.

7. Creating a Supplier Services SLAM Chart for reporting on the performance of shared Supplier Services into the System.

Individual & Team Accountabilty:

8. Verifying process accountability (Quality Assurance) at the end of each process value stream.

9. Collecting and storing Kata Card, Metric, and Training information in Historical Logs.

Corporate Liason:

10. Providing information to the group about market conditions, CCAPA updates, as needed, and other relevant corporate news every morning.

11. Creating a Cumulative System Performance Chart and being sure to have Process Owners Update it at the end of every Master Cycle. This chart is used for quarterly reporting purposes.

A Stable environment requires one or more Process Owners (PROs). Each PRO is responsible for the results of each process they own.

They work "in" the system, which includes:

Customer Mastery:

1. Ensuring that the current service levels represent their customer's current expectations.

2. Ensuring the process results (Quality Control) meet the current service levels.

3. Ensuring frequent and insightful communication exists between the customers and the Process Owner to be certain of #1 and #2. A byproduct of this is a high Customer Satisfaction Rating.

4. Establishing a healthy positive emotional relationship with their customers and suppliers. A byproduct of this is a high Net Promotor Score.

Supplier Mastery:

5. Ensuring inputs meet expected quality levels (Initial Quality Assessment).

6. Working with suppliers, indicating process and customer needs, to improve the material and information-based process inputs. A byproduct of this is improvements to process efficiency and effectiveness.

Process Mastery:

7. Ensuring the process, or Kata, is executed according to each Standard Operating Procedure and corresponding Kata Card (Quality Assurance). This is done by receiving the completed Kata Cards from the upstream process checkpoints, and handing them off, along with the newly completed Kata Card, to the downstream checkpoint, or to the Master Chief when at the end of the value stream.

8. Going straight to recovery models to recover systems and assets as quickly as possible. This minimizes downtime.

9. Asking coworkers for help, when needed. Brief Ad-hoc Kaizen Teams comprised of volunteers are extremely effective in overcoming sudden technical challenges.

10. Ensuring the Standard Operating Procedures, Kata Card templates, and recovery models remain updated and represent the best-known ways to perform and apply recovery steps to the process.

11. Ensuring newly discovered issues and recovery techniques are added to the Kata Card templates, or process or asset recovery models as new problems and root causes are encountered.

12. Updating Service Level Attainment Monitoring (SLAM) Charts each day.

If a team member is stumped with a technical challenge they are encouraged to ask for help. An ad hoc team called a Kaizen Team can be voluntarily formed to assist. Kaizen Teams should be comprised of volunteers. Companies that adopt Kaizen practices have learned the volunteer aspect of a Kaizen team is critical. They have learned that volunteers seem to generate the best ideas.

> Companies that adopt Kaizen practices have learned the volunteer aspect of a Kaizen team is critical. Volunteers seem to generate the best ideas.

Kaizen improvement efforts can be performed on several levels. They can be triggered at the daily Stand-up when a team member needs help, as described above. These are called Kaizen Blitz's. They can also be triggered in the Cycle Retrospective Meeting, when either the entire group identifies an item to improve, or one or more team members voluntarily identify an item in their work environment to improve. These are called Kaizen Projects.

The Three Domains

The three domains are shared by each role and include the Future, the Present, and the Past Domains. We must plan our work ahead of time before we can measure the results of our work. For this reason, in the next three chapters we'll discuss the Future, the Present, and the Past Domains, in that order.

The Future Domain contains information we'll collect and use later. The Present Domain contains our workflow. The Past Domain contains our history, and metric results.

All three domains are tied together through a performance console, which is how senior management is kept aware of present status.

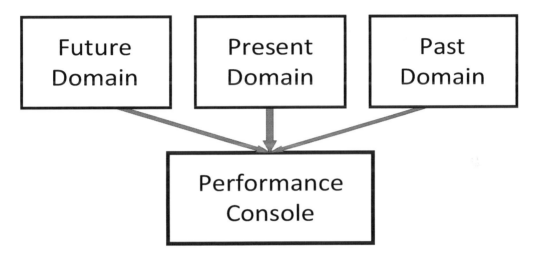

Figure 15 - The Three Domains of The Stable Framework™

The Four Core Meetings

The four core meetings that comprise a Master Cycle are the Cycle Planning Meeting, the Daily Kaizen Stand-up Meeting, the Cycle Review, and the Cycle Retrospective Meetings. These meetings are described in more detail in a later chapter.

The Five Values

Stable is about insight, coordination, and accountability within a system. These three elements bring results. Although it's natural for participants in a team to be primarily interested in their own accomplishments, one of the non-intuitive realities of teamwork is that the best way to maximize an individual's progress is to harness the power of the team.

It's important for team members to be accountable to each other. Good teamwork lowers stress and speeds up everyone's efforts. It increases communication which brings faster innovation and resolution to

> It's natural for participants in a team to be primarily interested in their own accomplishments. One of the non-intuitive realities of teamwork is that the best way to maximize an individual's progress is to harness the power of the team.

problems. Turnover diminishes, and productivity goes up, producing better overall results. The following values help teams perform at their best:

1. Respect People, Improve Systems.

Deming taught us that when problems occur, management must focus on improving the system and not on blaming the person. The old industrial model held employees accountable for problems and would often make them pay with their jobs. Deming taught us that the real question to ask is how the system can be improved so the problem does not occur again...both with the same employee and future employees working the same task. This "safe culture" engineering fosters trust and assurance among employees and drives creativity when problems occur. When a problem does occur, the Master Chief should challenge the Process Owner to brainstorm permanent improvements to the process to prevent those same problems from happening again.

For difficult problems, other members from the team or the entire team can be asked by a Process Owner to volunteer solutions and work together through the problem. Working together to solve problems in a short amount of time is encouraged. We call this a Kaizen Blitz. Blitz teams work best when they are comprised of members who have volunteered to help.

2. Speak with Data.

Whenever possible, use data to express reality. If you want to know how happy your customers are with your products or services, survey them. If you want to know how much time is spent reworking bad code that was rejected from the test department, measure it. The Stable Framework™ is all about improvement. You cannot conclusively improve something unless you first measure it.

Measurement has historically scared employees. In the past, measuring was used to reward or punish workers, and this was unfortunate. The proper application of measurement should be to provide a window into reality enabling employees and management to brainstorm ways to improve the system. This is an empowering experience and means that measurement should be a recurring and vocal part of any team's continuous-improvement process.

> The proper application of measurement should be to provide a window into reality enabling employees and management to brainstorm ways to improve the system.
>
> This is an empowering experience and means that measurement should be a recurring and vocal part of any team's continuous-improvement process.

As a group's objectives are made known to them, they should identify which drivers will bring success, and then which measures would indicate progress along those drivers. Measuring performance in this manner will keep the team unbiased and much closer to reality, meaning they will make better decisions along the way to sustained success. Measurement approaches should be scrutinized and revised regularly.

The team should always consider better ways of measuring to improve visibility into the reality of the status and improvements of its processes and process goals. Measurement should be discussed often and should be an ever-present part of the work culture.

Be sure to keep efforts to measure information in check with the value of the information.

Some measures we'll use may be components of larger measures and indicators of organizational progress. These are primary indicators and we call these measures Control Points.

3. Accountability is Binary.

Resist vague status updates like "we think we are on track," or "we feel like we are moving along OK." Comments like this are of little value. Instead, we work with small tasks and report the status as either done or not done.

Keep task sizes at two days or smaller wherever possible. Your team members should be constantly moving items through the Kanban and if they have tasks that are larger than one or two days the impression will emerge that nothing is getting done, and your teams may begin resenting the daily Kaizen Stand-up Meeting because they have no progress to report that is different from one day to the next.

In addition, Kata Cards submitted at process handoff checkpoints easily provide binary accountability. The only way to ensure accountability is by measurement.

4. Repetition with reflection and change brings improvement.

Repetition alone brings systematized inertia. Add reflection and you get better insight. Better insight generates new ideas and those new ideas requires change in something to bring improvement. Shewhart's model for gaining insight is the Plan-Do-Check-Act model. In this model, you make an improvement plan (Plan), then you act on the plan (Do), then you measure the results (Check), and if favorable, you update the Standard Operating Procedure and process Kata Cards to reflect the updated improvement (Act).

If the results are not favorable, of course, you'd skip the last step and try the next idea, working through the model.

In his popular book *Kaizen*, Masaaki Imai tells a story he heard from Toshiro Yamada, a Professor Emeritus of the Department of Engineering at Kyoto University. Yamada had visited an American steel manufacturing plant for a managerial discovery exchange when he was young, and then visited the same plant 25 years later.

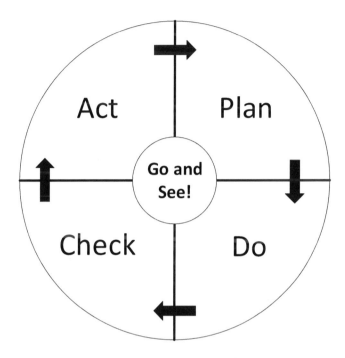

Figure 16 - Plan-Do-Check-Act

He was shocked to see no significant improvements in how the plant was managed. In 25 years, nobody had come up with any significant innovations or was permitted to try any major improvements. Don't let your environment be like this plant!

5. Everything can be Improved.

Stable practitioners operate under the belief that everything can be improved. Again and again, from Toyota to Ford to Hyundai and other companies that adopt Lean philosophies, these companies have realized that everything can be improved, and they operate with that mentality.

In Summary

Stable equips organizations with a concise, working model for insight, structure, stability, collaboration, accountability, and continuous improvement. All these elements implemented together create an Operational Excellence program for any organization.

Chapter 8 - The Future Domain

The Future Domain is where information is stored for future use. This includes our System Configuration, our System Schedule, and our System Backlog.

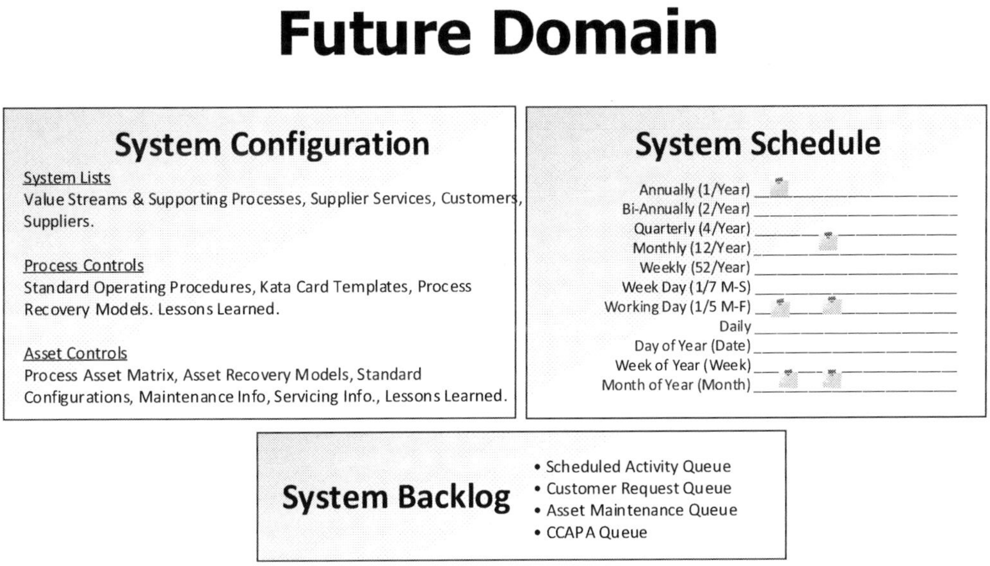

Future Domain

System Configuration

<u>System Lists</u>
Value Streams & Supporting Processes, Supplier Services, Customers, Suppliers.

<u>Process Controls</u>
Standard Operating Procedures, Kata Card Templates, Process Recovery Models. Lessons Learned.

<u>Asset Controls</u>
Process Asset Matrix, Asset Recovery Models, Standard Configurations, Maintenance Info, Servicing Info., Lessons Learned.

System Schedule

Annually (1/Year) _____
Bi-Annually (2/Year) _____
Quarterly (4/Year) _____
Monthly (12/Year) _____
Weekly (52/Year) _____
Week Day (1/7 M-S) _____
Working Day (1/5 M-F) _____
Daily _____
Day of Year (Date) _____
Week of Year (Week) _____
Month of Year (Month) _____

System Backlog

- Scheduled Activity Queue
- Customer Request Queue
- Asset Maintenance Queue
- CCAPA Queue

Figure 17 - The Future Domain

System Configuration

The System Configuration represents the institutional knowledge developed and stored for the retention and improvement of the environment. Using a Configuration Management System, System Lists, Process Controls, and Asset Controls can be stored and improved over time.

A good Configuration Management System is version controlled, backed-up regularly, is centrally accessible, and is easy to navigate.

Following is a description of each group of assets found in a Stable Configuration Management System:

System Lists

The System Lists are an assortment of any managed system information that should be kept in the environment. At minimum, this includes a list of the value streams (Master Services) and supporting processes, their customers, supplier services, and suppliers.

Value Streams and Master Services

The Master Services List is a collection of all the services offered by the environment. It should represent all the value propositions and corresponding supporting services. Think of this as the list of value streams the system provides.

Some of the services might be flagged as critical to the company's success. We call their metrics Control Points, and they are usually part of the overall company Key Performance Indicators (KPI's).

As part of Quality Planning, the Process Owners of each of the services should have an initial consultation with the major customers of each service and draft a specific service level for each service they are responsible for. The service level should be noted in the Master Services List. After the Process Owners do this, they need to report their findings to the Master Chief and demonstrate that the appropriate service level expectation has been noted in the Master Services List.

Master Service List		
System	SLA/SLG	Standard
Servicing ERP	Agree-ment	99.999% Uptime during business hours
Servicing Backup	Agree-ment	Backup Daily
Code Library Backup	Goal	Backup Daily
Daily Reconciliation Process	Agree-ment	SLA: Complete before 12:00 pm
Data Warehouse ETL	Agree-ment	SLA: Complete every 20 minutes

Figure 18 - Service Level Attainment Monitoring (SLAM) Chart

The status of these services is reported using SLAM Charts and Cumulative System Performance Charts, discussed in Chapter 10 – The Past Domain.

Optionally, the Master Services List could include pending and retired services, which are discussed next.

Pending Services

Pending Services are services that have not yet been deployed but are presently in development. This list is updated by the Master Chief from regular meetings with Engineering. It's important to keep this list updated so that no last-minute surprises impact an Operations group.

Problems in Operations begin in Engineering. Therefore, so do solutions. It's important the Ops, DevOps, Implementation, Support Desk, and even Sales, and Marketing leads should be attending a regular Engineering Status Meeting.

Often in software development, a well-intentioned but forgetful Engineering team will spend a year developing a new system on a new platform just to slide it under the door for Operations to begin supporting it, without any previous warning. Depending on the situation, Operations may be unequipped with the skills required to service the new offering, and chaos ensues. Keeping this list up to date is healthy.

Retired Services

Retired services are those legacy services that were once offered but have now been retired. Some customers may still use these, and they should stay on this list until there are no more legacy customers using them and they are not needed going forward.

Supplier Services List

Supplier Services are those incoming services required for your active services to function properly. These services are part of your supply chain.

If your Master Services require power, networking, a database backup, and an active internet connection, then you would list all of those in the Supplier Services List and report on their service level performance each day or at the end of your Master Cycle using a SLAM Chart. Obviously, your Master Services will be negatively impacted by your Supplier Service failures, so it is important to measure the performance of these also and present the finding to your suppliers if they are problematic.

Supplier Services List	
System	Service Level Agreement/Goal
Internet	SLA: 99.999% Uptime during business hours 100 MB Up, 50 MB Down
Data Center Power	SLA: 100% During business hours
Data Center Cooling	SLA: 100% During business hours 62% Server Room

Figure 19 - Supplier Services List

Customer List

The Customer List is where contact information for your primary customers reside. Customers should be associated with Master Services, listing primary customers and their contact information for each Master Service. It's a great place to note how often your customers expect to be surveyed for satisfaction ratings or ideas for product or service improvements.

Process Controls

The Master Services List is a list of processes organized by Value Streams. For each work process, a Standard Operating Procedure (SOP), a Kata Card template, a Processes Recovery Model, and a Lessons Learned document should be created.

The SOP can be a brief description, a diagram, a SIPOC Model, a video, or any other means of communication acceptable to the MC (Master Chief).

The Kata Card for the SOP is not a step-by-step instruction, but rather, a super checklist of lessons learned and risks to watch out for while performing the process. As we read in *The Checklist Manifesto*, written by Atul Gawande, checklists should be short. Several short checklists are better than one long checklist. Some checklists are meant to be completed in a step-by-step manner as the work is being performed, and some checklists are meant to be completed as the final step after completing work items. The Process Owner determines which type of checklist to build during Quality Planning.

> Some checklists are meant to be completed in a **step-by-step manner** as the work is being performed, and some checklists are meant to be completed as **the final step** after completing work items.

A Process Recovery Model is a tool used by Process Owners to quickly troubleshoot present problems based on past knowledge. As problems occur, Process Owners should be trained to immediately obtain the Process Recovery Model for that process and look at the list of past problems. If the problem is listed, they should look at the remediation steps and quickly act to resolve the problem. If the problem encountered is on the list it should be added once the problem has been resolved, so that future problems can be quickly remedied with help from past knowledge learned.

Asset Controls

The Configuration Management System includes an Asset Control section, listing all the primary assets required for the Process Owners services to function properly.

Assets should be linked to the services they support. This may be done with a Process Asset Matrix. A simple spreadsheet can facilitate this.

You may also want to track the last time the asset was serviced, or a recurring service schedule. For example, servers should be set to a recurring automated defragmentation schedule, and checked for CPU saturation, swap-file to RAM ratio, and disaster recovery planning every so often.

In software, the assets may be the database server, the file server, the application server, the web server, the internet connection, or any other standard asset the process needs to use to perform successfully.

Each asset should include an Asset Recovery Model, or ARM Sheet. This sheet contains the known past problems experienced with this asset, and the steps taken to recover from the known problems.

Like Process Recovery Models, Process Owners should be trained to go straight to this list when an asset fails to find out if the problem has a known cause already experienced in the past. They should be able to read the resolution quickly and get the asset back up and running.

It would also be useful to note any manufacturer warrantee information or support contact information for these assets, so that they can be quickly found during an emergency.

Any other Lessons Learned about these assets can be stored on a Lessons Learned document or field for each asset.

System Schedule

The System Schedule is a holding place for all recurring events impacting the system. The schedule is used to list future scheduled tasks which will then feed into the Present Domain's work queue, during the Master Cycle Planning Meetings.

It is organized by Yearly, Quarterly, Monthly, Weekly, Day of Week, and Daily categories. Following is an example of a System Schedule.

Looking at the previous figure, the notes indicate scheduled recurring items. These recurring items should be added to the Scheduled Work queue as needed.

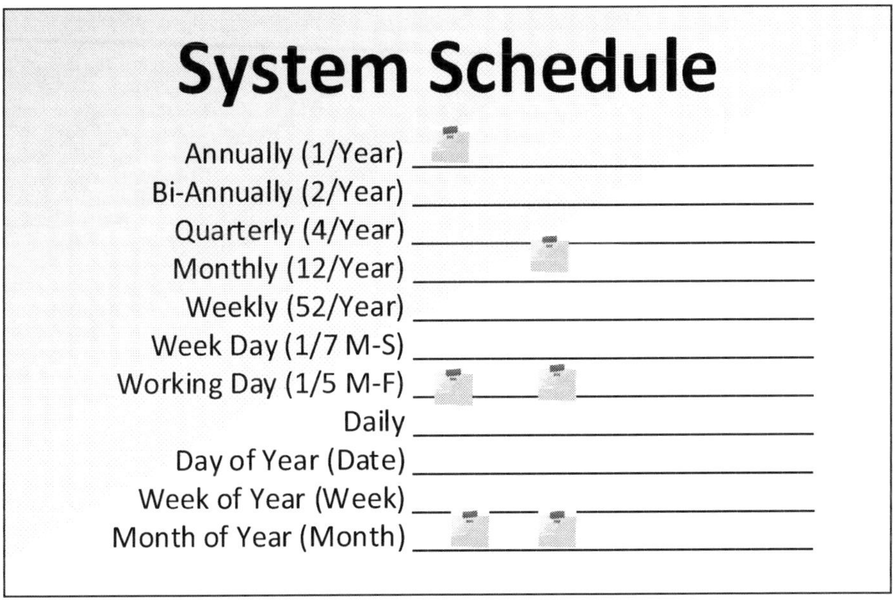

System Schedule

Annually (1/Year) _____

Bi-Annually (2/Year) _____

Quarterly (4/Year) _____

Monthly (12/Year) _____

Weekly (52/Year) _____

Week Day (1/7 M-S) _____

Working Day (1/5 M-F) _____

Daily _____

Day of Year (Date) _____

Week of Year (Week) _____

Month of Year (Month) _____

Figure 20 - Example System Schedule

System Backlog

The System Backlog represents all the incoming requests from various sources within and outside of the system. They are organized into the Scheduled Activity Queue, the Customer Request Queue, the Asset Maintenance Queue, and the CCAPA Queue. Each queue is described below:

Scheduled Activity Queue

Based on items listed in the System Schedule, upcoming work items that are time sensitive are placed in this queue to be worked on by the team in the Present Domain. The Master Chief is responsible for being sure time sensitive activities are added to the System Schedule and placed in the Scheduled Activity Queue as needed

Customer Request Queue

The Customer Request Queue is simply the bucket of urgent requests that materialize daily in any operations environment. In a small organization, these get handed to the engineering team, interrupting its daily planned activities. These constant interruptions are usually unavoidable and slow down scheduled project activities. In larger organizations, often a single engineer or an entire engineering team is devoted to handling these daily requests. In Stable, they are queued here and prioritized in real-time by the Master Chief to be worked on by the team in the Present Domain.

Asset Maintenance Queue

The Asset Maintenance Queue is the selection of service objectives that are regularly needed for equipment maintenance. In an IT environment this can be checking available hard drive space or CPU saturation while running critical services, verifying backup media, organizing CAT-6 wiring, or any other needed housekeeping. Think of the assets in your environment as a fleet of tools which require occasional maintenance.

CCAPA Queue

CCAPA is an acronym meaning Correction, Corrective Action, and Preventative Action. These are the remediation steps involved when unexpected production issues occur.

The CCAPA Queue comes from the Service Desk. When a customer calls a Service Desk with a nonconforming experience, we call that an Issue. When multiple customers call in about the same issue, we call that a Problem. The Service Desk should prioritize the problems and present an escalated Problem List to Engineering regularly containing the most critical problems to be solved.

This can be done daily, in an early morning meeting with the Master Chief just before the team's Kaizen Stand-up Meeting, or in a larger company moving slower, it can be done at a separate meeting once per week, per cycle, or during the Cycle Review Meeting.

For each incident reported, a Correction should occur for that customer, meaning a workaround, or replacement, or fix should happen. Next, the Process Owner should be tasked with identifying the root cause for the nonconforming product or service so that Corrective Action can be taken.

Corrective Action means that the Standard Operating Procedure, and Kata Card for the process that failed should be improved to prevent that problem from recurring.

Finally, Preventative Action means that any additional processes that might contain the same vulnerability should also be improved. The status of the nonconforming problems in the CCAPA Queue should be managed between the Service Desk and the Process Owners using Vertical Communication.

Chapter 9 - The Present Domain

The Present Domain is where work gets performed. The team uses a visual workflow tool called a Kanban Board to indicate what they are working on and the completion status of that work.

Present Domain

Team
- Daily Kaizen Standup
- S5 Housekeeping
- Adhoc Kaizen Teams
- Process Kata & Kata Cards
- Training/ReTraining

Committees
- Change Control Committee
- CCAPA Committee

Environment
- Alert System
- Production Change Log
- Suggestion Box

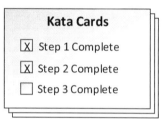

Figure 21 - The Present Domain

The Kanban Board

A Kanban is a visual tool for managing workflow. It is a simple work authorization system that allows the team to respect natural work-in-process limitations inherent in the environment. It also keeps the task queue upstream, where it belongs.

The capacity in every environment is naturally limited by the number of people available to perform the work. Whether working on formal project tasks, or reactive emergency request tasks, a Kanban matches tasks with available people, thereby limiting allocated tasks to only the people available to work on those tasks. Consistent with the One-piece flow process design, this eliminates excessive task switching which causes unnecessary waste and provides no value to the customer.

The basic columns on a Kanban board are organized by "Objectives," "Selected Tasks To Do," "In Process," and "Done." The "Objectives" column is a list of items sourced from the System Backlog which is a combination of the Customer Request Queue, the Scheduled Activity Queue, the Asset Maintenance Queue, and the CCAPA Queue. The Master Chief owns the triage process and prioritization of items from these queues into the "Objectives" column. The team of Process Owners, as they finish past tasks and become available for more work, will then self-select items from the "Objectives" column and move them into the "Selected Tasks To Do" column. As they receive requests from their customers during the day, they can add them to the "In Process" column and finally to the "Done" column when complete. When they are freed up for more work, they can go back to the Kanban board and select the next item from the top of the "Objectives" column.

Some groups add an "On Hold" column for blocked tasks that are delayed for substantial amounts of time. Items in this column might require escalation to higher authority to be expedited.

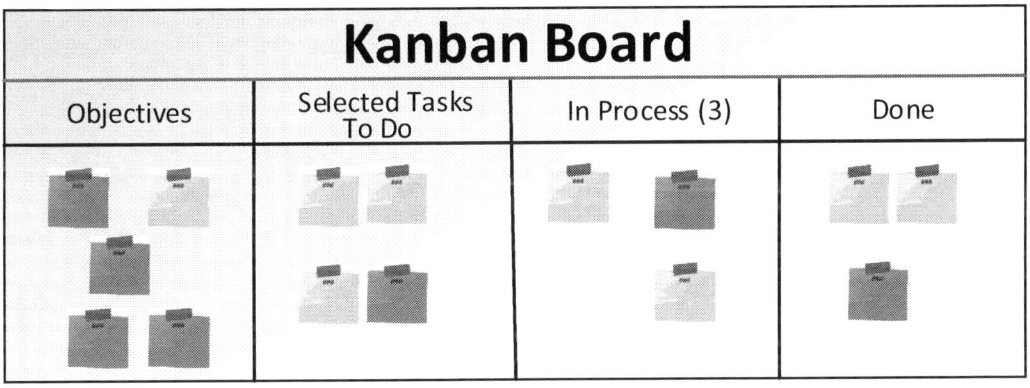

Figure 22 - Kanban Board

Project work can be completed using a Kanban process, if wanted. The project team can simply break down their projects into multiple work packages of user stories or use cases, and then decompose those into individual activities to be completed. At the beginning of each Master Cycle, the team can estimate how many activities can be completed during the cycle and put them into the "Selected Tasks To Do" column.

Some groups create horizontal swim lanes on their Kanban Board and use one lane for each project. Another technique is to use different colored sticky notes for different projects. We've learned in books like *The Phoenix Project*, and *The Goal*, that task switching creates an illusion of more work getting done but actually slows down the flow of value within a system. For this reason, the "In Process" column contains a work-in-process limit. In the example below, it is 3. This means no more than three tasks should be in the "In Process" column. Protecting natural bandwidth limitations reduces Work-In-Process (WIP) inventory and keeps us in One-piece flow mode. Otherwise, our bottleneck would start collecting WIP inventory which hurts the entire value stream.

With that knowledge, a Kanban tool enables the team to stay optimized to deliver value as fast as possible.

Daily Kaizen Stand-up

Every morning, the Master Chief and the Process Owners meet together in a morning Kaizen Stand-up Meeting to hear new suggestions from the suggestion box, to get an optional market summary report from the Master Chief, and to discuss the daily progress and objectives for which the team is focused.

The Master Chief puts the highest priority objectives in the "Objectives" column of the Kanban board, in order of most to least urgent. The team can pick from there when loading new work into the "Selected Tasks To Do" column of the team's Kanban flow.

> Wherever possible, the Process Owners should be cross trained to work on each other's processes and tasks. In Lean, we say **it's the goal, not the role**, that is important.

As work items are completed, their status moves through the Kanban from "Selected Tasks To Do," to "In Process," to "Done." The team may rename the individual columns in their Kanban board to best suit their environment.

When the items are prioritized by the Master Chief in the Kanban and put in the order of what is needed most, this is called a Pull System. The Process Owners can view the "Objectives" column and select the most urgent objectives they are qualified to complete.

Wherever possible, the Process Owners should be cross-trained to work on each other's processes and tasks. This eliminates the risk of nobody being able to perform the task because that one person who knows how to may not be in the office today. SOP's and Kata Cards make this easier. In Lean, we say it is the goal, not the role, that is important.

Measuring Velocity with a Kanban Board

Work Items should be categorized and time stamped as they move from column to column on the Kanban board so the group can get accurate historical information about the quantity of and turnaround times that different types of work items take to move through the team's Kanban. This way, empirical work completion times for each category of work items can be collected and tabulated over time. This gives teams better ways to measure how much work is being performed within each Master Cycle, and how long a typical item from a specific category will take to move through the Master Cycle, under typical environmental conditions.

If your team performs both project and operations work, and the project work requires estimation and tracking over time, it could simply be estimated up front using effort points, or Stones, and tracked for velocity per Master Cycle over time. Each objective could be time stamped for when it entered the Kanban, and then when it was completed. This helps with estimates of similar sized objectives in the future.

In this manner project work and Operations tasks combine on the same Kanban Board. You can measure project work velocity using user story points, or stones as we call them in Stable.

Operations tasks are easier to account for by categorizing them into types of requests, and then time-stamping their start-times and end-times through your system. Then you can report on how many of what types of requests have made it through your Kanban during each Master Cycle, and the average length of time required to complete each task.

Tracking this kind of information becomes valuable for setting future customer expectations correctly.

Process Kata and Kata Cards

Executing and completing work in a repeatable process is known as performing a Process Kata. A systemized process should have a Standard Operating Procedure created for that process and a Kata Card Template created to be completed while the Process Owner performs the kata. The completed Kata Card is then turned in to the next internal downstream customer at the next checkpoint. If there is no next internal customer, then the Master Chief gets the final stack of Kata Cards for that process flow, and that is the final checkpoint for that process value stream.

Kata Cards are a kind of super-checklist. They are a risk-management tool ensuring everything was performed properly, and they may contain process data as needed. Think of completed Kata Cards as certificates that the work was performed correctly and completely.

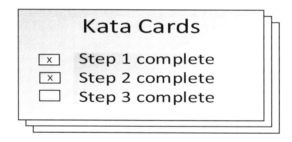

Figure 23 - Process Kata Cards

The Master Chief stores the accumulating stacks of Kata Cards in a Kata Historical Repository, where a specific completed Kata Card can be retrieved if needed during an audit or a CCAPA event.

> Think of these completed Kata Cards as **certificates** that the work was performed correctly and completely.

The full cycle of the Process Kata involves knowing the customers, establishing a service level agreement with them, or just a service level goal with the Master Chief if a service level agreement is not appropriate, and working through the process using a Kata Card always looking for ways to improve the process. In some situations, an "As Is" and a "To Be" analysis may be created with improvement plans made to achieve the desired service levels.

Although completed Kata Cards at the end of a process may be handed to the Master Chief at any time, the most convenient time is during the morning Kaizen Meeting.

Ad-Hoc Kaizen Teams

Cooperation is encouraged in Stable. Occasionally, a Process Owner may get stuck on a challenging activity and need more ideas or different approaches. During a morning Kaizen Meeting, or anytime during the day, the Process Owner can request group help and form a Kaizen Team to help get past the challenge. Anyone with available time can volunteer to assist in this ad-hoc manner. We call this a Kaizen Blitz.

CCAPA Committee

The CCAPA Committee owns the CCAPA list and the committee meeting is attended by the Service Desk Director, the Development Director, and the Master Chief. This meeting can happen daily, before the morning Kaizen Stand-up Meeting, or once per cycle, preferably during the Cycle Review. The purpose of the meeting is to be sure all parties are aware of the status of any new or outstanding non-conforming issues. New issues can be brought to the attention of the team by the Master Chief during the morning Kaizen Stand-up meeting.

Change Control Committee

The Change Control Committee is the whole Process Owner team and any other external members who are required to approve a change. This change control process can be daily, during the morning Kaizen, or once per cycle during the Cycle Review Meeting. If changes created from this group need to flow downstream to a larger body, the committee can approve changes internally during morning Kaizen, then the approved change documents can be sent to the larger multi-department Change Control Meeting which would meet at regularly scheduled intervals.

Training, Retraining, and Cross-training

Formal quality programs must prove that all employees have received training and are aware of their work processes. This means there is a record of their having read the Standard Operating Procedures required to perform their jobs.

During Quality Planning, when the SOPs are created by the Process Owners, this represents their first completion of training and can be logged as such. In practice, reading an SOP is never popular, therefore retraining should be scheduled perhaps every six months, or annually, where the Process Owner in charge of each process reviews all the SOP's governing his or her processes and makes sure they are up-to-date. The SOP's should reflect the best-known ways to perform the work. These SOP's can be brought up to date every six months or annually, during Quality Planning.

If the team finds a better way to perform a process, they can update the SOP at any time. The SOP, and all the process controls should be stored as successive versions under document change control.

A cross-trained team is a good idea in Stable. If a team member is absent another team member can perform a critical task for them. A Training Log should show that every Process Owner has been trained on every other Process Owner's processes, where appropriate.

When a nonconforming issue is experienced by a customer, one of three things should happen. Either the process needs to be improved internally, or it needs to be protected against some outside factor that impacted it, or the Process Owner requires retraining to perform the process correctly going forward. Frankly, this last option means the operator didn't follow the process, regardless of whether there was a record showing that they did or not. In situations like this, where the employee was confused, re-training and logging the re-training is appropriate. A training log should be kept by the Master Chief showing when everyone was trained, and when they are queued up for retraining. Scheduled training should be listed in the System Schedule.

5S Housekeeping

A clean, orderly environment is a critical part of value flow in a workplace. Toyota discovered its employees could work with clear minds and better concentration when they invested just a few minutes each day performing housekeeping activities.

Every morning after the daily Kaizen Stand-up, the team returns to their work environment and performs 10 minutes of cleanup. This includes tidying up the immediate environment, and can also include updating any SOP's, Kata Cards, customer lists, or other process controls.

An excellent technique for workplace housekeeping is a process called 5S. 5S is an organizational philosophy made famous by Just in Time manufacturing practices. It represents five organizational concepts that begin with the letter "S." Adjusted for IT, they are:

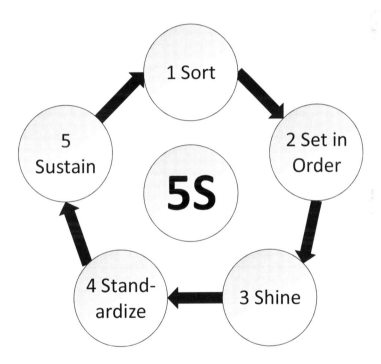

Figure 24 - 5S Housekeeping

1. Sort

1. Dispose of any items not necessary in the workplace. This includes items on walls, bulletin boards, aisles, and stairways.
2. Reducing the number of items in the workplace is always good.

The goal is to have only those items needed in the workplace that are required to perform the work.

2. Set in Order

1. Arrange items so they can be found when needed, by creating a place for everything and then putting everything where it belongs.
2. Examples: Sticky Notes, Projector Remote, Phone Numbers, Whiteboard Markers, etc.

3. Shine

1. Clean your workplace daily. Be sure equipment, floors, walls, stairs, and surfaces are free from dirt and grime. Clean work areas promote better and faster decision making.
2. Be sure labels, signs, and so on are clear and obvious.
3. Be sure cleaning materials are easily accessible.

4. Standardize

1. Create and follow a set of regular procedures to keep the work area orderly and clean.

5. Sustain

1. Are the 5S rules being practiced daily? Have all employees received 5S training?
2. Stay on top of things using 5S principles, so the environment doesn't deteriorate.

Alert System

Part of any operational environment is some type of alert system that should provide real-time status updates regarding environmental assets and their attributes.

This can take many forms, and some readymade tools exist for this purpose. You can use a network monitoring tool or create your own alert system to send email alerts or text messages when something is not as it should be.

Scoreboards are becoming popular today, where a large screen is placed in a work area showing the real-time status of the environment.

These tools equip teams to be constantly aware of performance status and be quick to respond to problems.

Production Change Log

If applicable, some groups keep a Production Change Log. This makes it easy to determine what changed in the environment in the past on a certain date when troubleshooting production problems.

Suggestion Box

A suggestion box should be clearly visible and easy to access. Suggestions should be part of the team's culture. Each morning the Master Chief should check the suggestion box and read the contents to the team. A twitter feed could also be used for this.

Suggestions should be taken seriously, and the Master Chief should provide an explanation for any suggestion not able to be implemented.

One indication of the health level of team morale is the number of suggestions that are submitted by the team over time. Disempowered teams don't make suggestions. Empowered teams do.

Chapter 10 - The Past Domain

The Past Domain contains historical information regarding the most recent and trending status reports from the environment. Charts created and used for reporting here will be posted on the Performance Console.

There are three types of data in the Past Domain: Master Cycle Totals, Historical Logs, and a Cumulative System Performance chart. This latter chart is the quantitative summary of Key Performance Indicators that can be brought to and discussed at the organizations Quarterly Business Reviews (QBR's), if needed.

Master Cycle Totals

Each of the Process Owners report on the daily performance of their Master Service and Supplier Service Value Streams using Service Level Attainment Monitoring (SLAM) Charts. In addition, a single Supplier Service SLAM Chart is updated by the Master Chief for all shared Supplier Services.

Past Domain

Master Cycle Totals
- Cumulative System Perf. (CSP) Chart
- Supplier Services SLAM Chart
- Cycle Totals
- Customer Satisfaction
- Customer Net Promoter Score
- SLAM Chart A, B, etc.

Historical Logs
- Production Change Log
- Critical Decision Log
- Kata Card Historical Repository
- CSP Repository
- Training Log

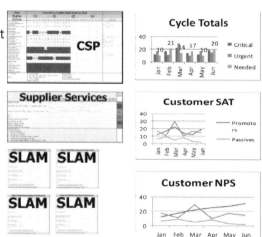

Figure 25 - The Past Domain

In addition, Master Cycle metrics such as total and type of work items completed, customer satisfaction ratings, customer Net Promoter Score ratings and any other important metrics are reported. SLAM Charts should show target service levels vs. actual service levels attained.

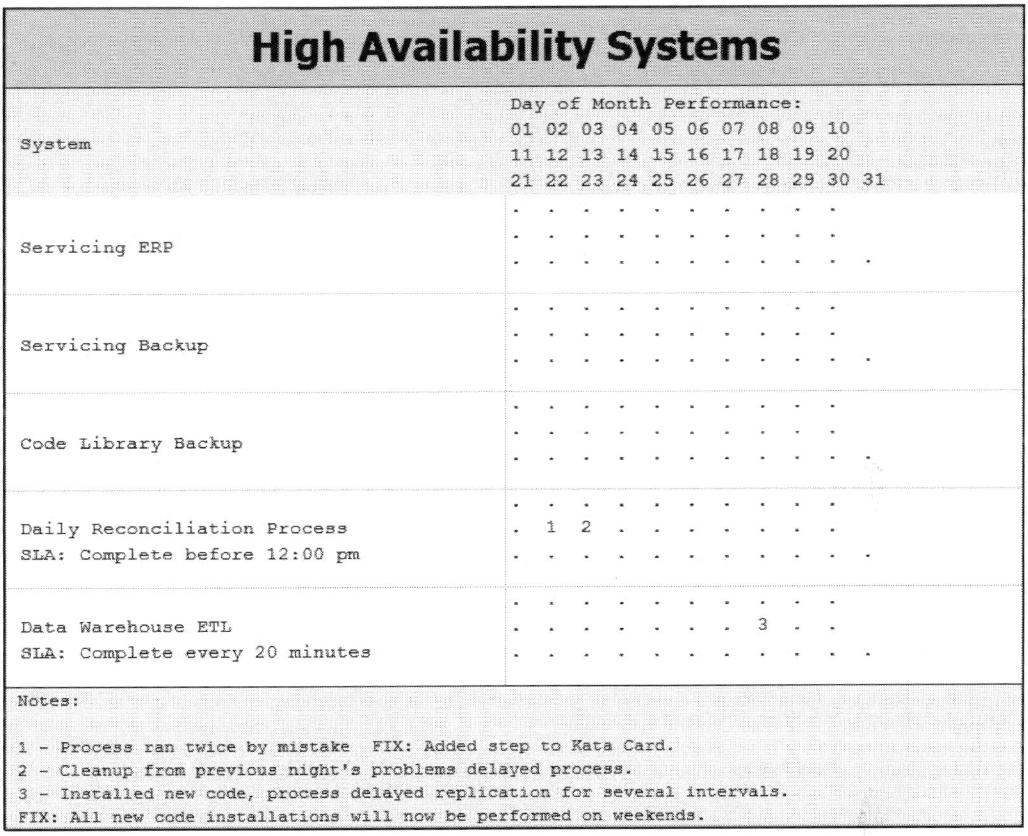

High Availability Systems

System	Day of Month Performance: 01 02 03 04 05 06 07 08 09 10 11 12 13 14 15 16 17 18 19 20 21 22 23 24 25 26 27 28 29 30 31
Servicing ERP	. . . .
Servicing Backup	. . . .
Code Library Backup	. . . .
Daily Reconciliation Process SLA: Complete before 12:00 pm	. . . . 1 2 . . . .
Data Warehouse ETL SLA: Complete every 20 minutes	. . . . 3 . . . .

Notes:

1 – Process ran twice by mistake FIX: Added step to Kata Card.
2 – Cleanup from previous night's problems delayed process.
3 – Installed new code, process delayed replication for several intervals.
FIX: All new code installations will now be performed on weekends.

Figure 26 - A SLAM (Service Level Attainment Monitoring) Chart

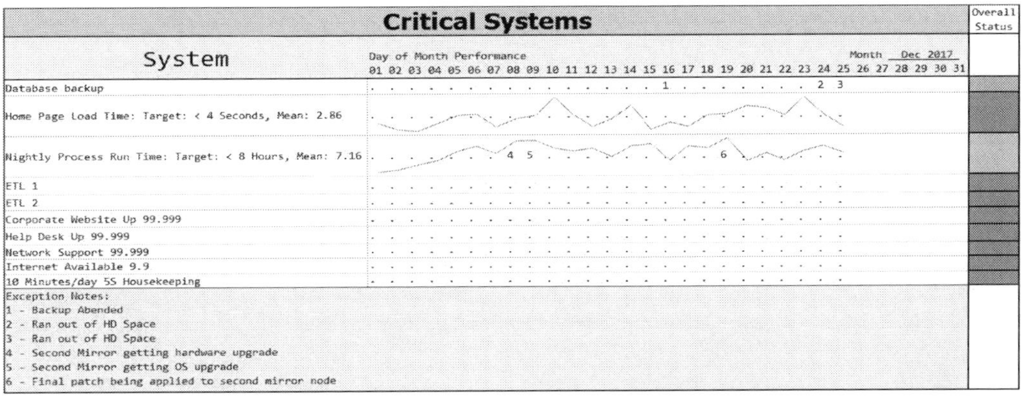

Critical Systems		Overall Status
System	Day of Month Performance Month Dec 2017 01 02 03 04 05 06 07 08 09 10 11 12 13 14 15 16 17 18 19 20 21 22 23 24 25 26 27 28 29 30 31	
Database backup	. . . . 1 . . . . . . 2 3	
Home Page Load Time: Target: < 4 Seconds, Mean: 2.86		
Nightly Process Run Time: Target: < 8 Hours, Mean: 7.16	. . . . . 4 5 . . . . 6 . .	
ETL 1	. . . .	
ETL 2	. . . .	
Corporate Website Up 99.999	. . . .	
Help Desk Up 99.999	. . . .	
Network Support 99.999	. . . .	
Internet Available 9.9	. . . .	
10 Minutes/day 5S Housekeeping	. . . .	
Exception Notes: 1 – Backup Abended 2 – Ran out of HD Space 3 – Ran out of HD Space 4 – Second Mirror getting hardware upgrade 5 – Second Mirror getting OS upgrade 6 – Final patch being applied to second mirror node		

Figure 27 - Advanced SLAM Chart Containing Statistical Process Control Data

A SLAM Chart may be as simple as indicating that each service level was met each day for all services. Or, they may contain other useful information such as average usage rates, average times, customer satisfaction rates, etc. The two preceding charts are examples of a simple, and then a more complex SLAM Chart, respectively. Notice the Statistical Process Control Data showing on the second chart. Also note the solid boxes in the right-most column. These boxes would be colored green, yellow, or red, depending on the status of the service so far, during the Master Cycle.

Cycle totals include a profile of the types and amounts of tasks completed by the team during the past Master Cycle. Tasks can be organized into any group categories such as Critical Fix, Pre-Emptive Fix, Standard Work; or Internal Customer, External Customer; or perhaps by Sources of requests.

Easy ways to collect customer satisfaction ratings include surveys at the point of contact with each customer, Service Desk surveys, or random customer outreach surveys.

Project Work

Project status in Stable is reported on using a Master Project Schedule. This schedule shows the summary status of any projects being worked on in a Stable environment. For an example, please see the "Stable Portfolio Management" chapter in this book.

If detailed status information is needed for a particular project, a Gantt Chart and a Burnup Chart, and sometimes a Work Breakdown Structure is used. Please see the "Stable Projects" chapter for an example of these charts.

Stable uses a hybrid approach to managing projects that incorporates the best parts of traditional Plan-based and Agile project management. This gives the team the benefit of managing project dependencies using a Gantt Chart, and managing work items using a Kanban Board and a Burnup Chart. Please see the chapter titled "Stable Projects" for more information.

Non-Project Work

Non-project activities can be tracked several ways. If helpful, a chart showing the total number of tasks completed per cycle, by category, can be created.

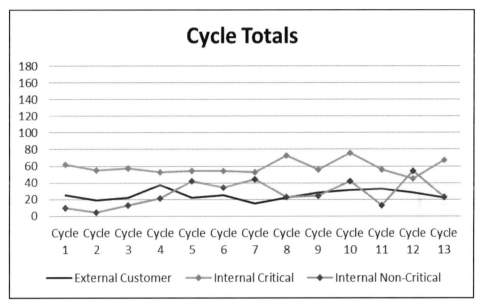

Figure 28 - Cycle Totals Showing the Number of Items Completed per Master Cycle

If the individual tasks are timestamped upon arrival, and as they are completed within the Kanban, averages can be collected over time showing the typical historical timeframe for fulfilling certain kinds of requests.

This can help manage customers' expectations better by providing a probable completion time for similar request types going forward.

Another helpful chart may be a Workload Analysis pareto chart showing the hours spent per request type, during the past cycle. This may be useful for brainstorming innovative ways to empower the customers or apply automation to reduce the burden on the Operations, or DevOps groups.

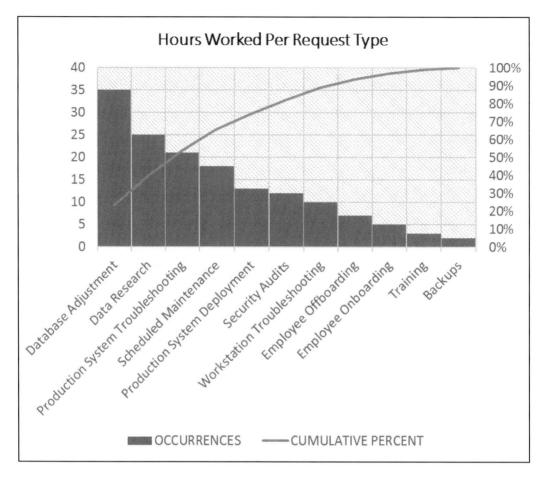

Figure 29 - A Workload Analysis Pareto Chart

Cumulative System Performance

As your group works through Master Cycles during the year, you will want to understand performance trends over time for your System. Most companies review performance quarterly. For this reason, we create a Cumulative System Performance Chart showing Master Cycle trends over time. This chart is created by the Master Chief but is updated during each Cycle Review Meeting by each Process Owner.

Cumulative System Performance 2018

Year Quarter M.Cycle	Q1						Q2						Q3			Q4											Trendlines
	1	2	3	4	5	6	7	8	9	10	11	12	13	14	15	16	17	18	19	20	21	22	23	24	25	26	
Master Services																											
• Executive Sponsor (Customer Satisfaction 1-10*)	3	7	9	10	8	7	6	7	8	9	10	8	9	10	10												
• Team Morale (Customer Satisfaction, 1-10*)	10	10	8	10	7	5	9	10	9	10	10	10	10	10	10												
• Servicing ERP (SLA/SLG MET? R/Y/G*)	#1			#4	#5	#6					#9																
• Servicing ERP (Rework/Scrap Hours, 0*)	12	45	15	13	14	5	8	6	10	9	12	6	12	10													
• Servicing ERP (Total Hours, Low*)	20	55	23	21	20	19	21	22	19	18	19	20	21	22	21												
• Servicing ERP (Cust Satisfaction 1-10*)	10	1	10	10	5	4	4	8	10	10	9	9	10	10	10												
• Servicing Backup (Net Promoter Score, 10*)	7	6	7	7	8	7	7	8	7	8	9	6	7	8	9												
• Servicing Backup (SLA/SLG MET? R/Y/G*)						#7																					
• Code Library Backup (SLA/SLG MET? R/Y/G*)																											
• Daily Recon Process (Total Hours, Low*)	10	0	4	12	5	7	8	7	4	8	4	5	9	7	6												
• Data Warehouse ETL (SLA/SLG MET? R/Y/G*)			#2	#3						#8																	
Supplier Services																											
• Internet (SLA/SLG MET? R/Y/G*)																											
• Network Support (SLA/SLG MET? R/Y/G*)																											
• DB Backup (SLA/SLG MET? R/Y/G*)																											

* = Desired

#Notes:

#1 Power Outage	#6 Power Outages
#2 New Division onboarding slowed process	#7 Network Upgrade
#3 Verification Failures	#8 System Upgrade
#4 Power Outages	#9 Backup Failed, Cluster FailedOver
#5 Power Outages	

Figure 30 - Cumulative System Performance Chart

The previous diagram shows an example of this chart. Note that it contains Executive Sponsor and Team Morale satisfaction scores, and then Master Services and Supplier Services metrics.

Note the trendlines along the right side, and the descriptions of deficiencies along the bottom. Remember that this data is collected to bring the group closer to reality, and to trigger constructive dialog on how the system can be improved going forward. Remember the two System Vectors: If customer satisfaction is down, focus on improving that first. Once that is where it needs to be, then focus on reducing time and costs by reducing rework, scrap, and recovery time.

A Cumulative System Performance Chart would be brought by the Master Chief to the Quarterly Business Review meeting, where the Master Chief would report on trends in his or her area, and set goals and create initiatives for wide-scale improvement.

Historical Logs

Historical logs include a Production Change Log, a Critical Decision Log, a Training Log, a past Kata Card Repository, a past SLAM Chart Repository, and a Cumulative System Performance Chart Repository.

Storing past Kata Cards and SLAM Charts are important for historical reference purposes when past quality information is needed.

A Production Change Log is an excellent tool uncovering root causes of issues in a production environment. When an anomaly is detected, it's helpful to match the start of the anomaly to a past change entered in the log.

A Critical Decision Log can be helpful for tracking the history of important environment decisions, or process decisions.

The Cumulative System Performance Log is the summary of improvement within the system over time. This is the summary of value the System has produced for the organization.

The Performance Console: Tying All Three Domains Together

All these domains are tied together using a Performance Console. The Performance console is where you post all the visual information summarizing the value streams.

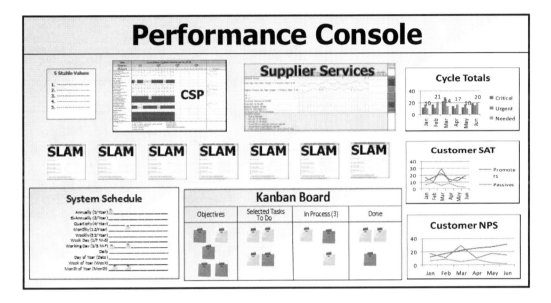

Figure 31 - The Performance Console

The performance console should be a wall space dedicated to tracking the group's performance. In Lean, we call this technique Visible Management. In Agile, we call this space an Information Radiator. According to Alistair Cockburn, a good Information Radiator has the following characteristics:

- It's large and easily visible to the casual, interested observer.
- It's understood at a glance.
- It is easily kept up to date.
- Information on it changes periodically so that it is worth visiting.

The Performance Console is the meeting place for the team. Morning Kaizen Stand-up Meetings happen in front of the console. Virtual teams will need to use a webcam or special software tool to connect the team with this information.

The goal of the information posted on the Performance Console is to present the overall performance of the department in a manner so clear that a person walking in off the street could see this information and know what is going on in the department. If a person walking in off the street can read and comprehend the information, then so can senior management.

This enables senior management to be freed up to spend their efforts growing the business instead of having to chase the group around probing for answers. We call this level of reporting Operational Excellence.

Pay attention to any other information senior management asks about and add it to the Performance Console. This increases your team's value and contribution to the business-IT alignment within the organization.

Chapter 11 - The Four Core Meetings in Stable

The Stable Framework™ includes four core meetings:

1. The Cycle Planning Meeting which occurs at the start of each Master Cycle.
2. The daily Kaizen Stand-up Meeting.
3. The Cycle Review Meeting where progress is reported to stakeholders.
4. The Cycle Retrospective Meeting where the team reflects on how to improve their environment.

These meetings repeat with each Master Cycle. In addition, two optional but recommended meetings are the Change Control Committee and the CCAPA Committee Meetings, which were introduced in Chapter 9 and are discussed more in Chapter 14.

Many organizations also hold a Quarterly Business Review meeting. The Master Chief may attend those meetings and present the Cumulative System Performance Chart.

The Master Chief must also be aware of the status of any soon-to-be completed Engineering projects. The Master Chief can be updated on this information by attending Development status meetings, or the Master Chief can collect this information through vertical communication.

The Master Cycle Planning Meeting

The Master Cycle Planning Meeting is called by the Master Chief and is held at the start of each Master Cycle. In this meeting objectives are selected from each of the work queues as needed and added to the Objectives column of the Kanban board to be completed as individuals on the team are freed up to work on more items.

The Master Chief orders the list of Objectives from top to bottom and the team self-selects which work objectives to do from the top of the list. Process Owners on the team self-select who gets the next item on the list.

During this meeting, objectives are broken down into individual tasks required to complete each objective. Tasks should be kept to one or two-day sized units of work. Expected one-or-two day time frames are then assigned to each task by the person who selected the task to complete. Most objectives should be familiar work, so the breakdown should already be known.

Process Owners should add any time-triggered items they need to the System Schedule, to be done later during planned intervals.

Project work is covered in chapter 15, "The Stable, or 'Hybrid' Project." Some groups create horizontal swim lanes, or use different colored sticky notes to separate multiple projects being worked on the same Kanban Board.

The Daily Kaizen Stand-up Meeting

The Daily Kaizen Stand-up Meeting is called by the Master Chief. The meeting should last no longer than 25 minutes. The meeting consists of the following:

- The Master Chief provides a brief, 5-minute or less market status report, which is a summary of any outstanding news from Senior Management or the Service Desk or any other part of the company. This information comes from a daily CCAPA Committee meeting, or from daily conversations with the Service Desk. If daily seems to frequent, the Master Chief may want to do this weekly.
- The Master Chief notifies the team of any new items or items that have received a higher priority.
- The team updates their SLAM Charts with yesterday's information and reports on the following four questions:

  1. What did I accomplish yesterday (including any change control items)?
  2. What do I intend to accomplish today?
  3. What roadblocks are in my way?
  4. What did I improve in my environment since yesterday?

This Kaizen Meeting is a daily event where the whole team can discuss and inform each other concerning what each Process Owner has completed and what they will be working on next. If a new high-priority objective has appeared in a work queue the Master Chief can add it to the top of the "Objectives" column on the Kanban board and present it to the group and discuss it briefly.

During this meeting each team member is reporting to the team. Once the team starts reporting, the Master Chief should do what they can to not be the focus of the dialog. Two excellent techniques are to remind the group that this second part of the meeting is for the team to report to themselves, and then walk behind the circle breaking eye contact.

Immediately after the stand-up meeting has ended, the team returns to their workstations and performs 10 minutes of 5S improvement. A clean environment enables faster and better decision making, so this activity is performed daily.

The Master Cycle Review Meeting

At the end of the Master Cycle, the Master Chief calls the Master Cycle Review Meeting. In this meeting, the group meets to showcase their work performance during the past cycle and discuss recent customer feedback. Implemented suggestions are explained. Total work completed is reported, the Cumulative System Performance Chart is updated, and old SLAM charts are filed away historically.

This meeting is called by the Master Chief and attended by the team, and any customers invited by the Master Chief and Process Owners.

The Process Owners are tasked with collecting customer feedback to bring to these meetings. It's important for Process Owners to be aware of their current customer satisfaction ratings. It may be useful for the Process Owners to set up a regular performance review survey with their primary customers and report on their findings. They can also work with the Service Desk for customer satisfaction information.

Kaizen Project results, which are described in the next section, can also be discussed in the Master Cycle Review Meeting, if appropriate.

The Master Cycle Retrospective Meeting

The Master Cycle Retrospective Meeting is the final meeting in the Master Cycle. It is called by the Master Chief and attended by the team, but not any customers. In this meeting the team discusses the following:

1. What went well this past cycle?
2. What could have gone better this past cycle?
3. What still feeds the Hidden Factory?

If no feedback or discussion is happening, it may be that the Master Chief needs to be the topic of discussion. The Master Chief may consider asking the team to self-select a scribe to write down comments the team can discuss and leave on the table for the Master Chief to pick up after the meeting has ended.

Kaizen Projects

In addition, the team should brainstorm and identify one or more target items to improve for the next cycle. The group can either brainstorm together and come up with an idea, or individuals can provide their own ideas to work on during the next cycle. These items should be something the Process Owners want to have improved.

Examples include automating an existing process, reorganizing an existing workflow, creating a customer education asset, or creating a new customer survey.

A team-based Root Cause Analysis can be performed using an Ishikawa Diagram to identify problem areas collectively. The team can then use their conclusions to identify a group based or individual Kaizen Project. See the chapter titled "Data Collection and Analysis Tools" for details about using an Ishikawa Diagram for this.

Chapter 12 - The Process Owner

The Process Owner, or PRO, is an intelligent, capable decision maker, who is singularly responsible for the success of each process he or she performs and the satisfaction levels and Net Promoter Scores of the PRO's customers. Capable means the PRO is empowered to make decisions, is aware of the organization's intent, and can interpret reality correctly. Since the PRO is the single person responsible for the success of the processes for which he or she owns, it's imperative that the PRO has a healthy relationship with his or her suppliers, understands all the components of value flow within his or her processes, and knows the expectations and needs of his or her customers.

Many of the PRO techniques discussed in this section are life skills, and in addition to improving workplace performance, can help the PRO improve many aspects of his or her own life. These techniques include identifying and engaging customers, making better decisions by seeking better data, understanding the root causes of problems, and progressing using continuous improvement techniques. All these skills will enable better long-term results in a Process Owner's life both professionally, and personally.

Customer Engagement

Process Owners should regularly collect actional feedback from their customers. This should be qualitative information, not quantitative information. Process Owners should continually be asking, "Is there anything else you need?" or "Can our service be improved in any way?" This can happen every time a Process Owner engages a customer.

Customer satisfaction metrics, usually collected in quantitative form, should be collected by surveys, questionnaires, or direct contact from the Master Chief. Customer Satisfaction information should never be solicited by the person responsible for customer satisfaction. This will bring skewed data. I remember quite being told by a restaurant host if I gave him 5 stars on his survey he had just handed me I would receive a free milkshake. I now call this the "milkshake problem."

In some cases, customer feedback can be coordinated through

marketing, the Service Desk, or a centralized customer feedback effort within the company.

It's important for the Process Owner to understand he or she has more than one customer. In fact, the Process Owner has many customers. Four types of customers include:

1. The business consumer at the end of the value stream.
2. The worker at the very next step in the process who receives the completed work item.
3. Another worker skilled in the art who may take over the process as the company grows, or when staffing is low.
4. An auditor in a regulated industry who will want to see proof of process consistency, or an internal auditor checking for compliance.

To summarize, actionable feedback should come from the Process Owner, whereas customer satisfaction metrics should come from an automated survey, Marketing, Support Desk, or the Master Chief.

Process Measures & Service Levels

Part of the responsibility of owning a process is being able to report on its status and progress accurately. Only part of the responsibility of the Process Owner is to perform the work, the other part is to be able to express clearly how the work is progressing against established Service Levels.

At a minimum a Process Owner could report daily on whether the service was delivered at expected levels or not, using a SLAM Chart.

An advanced measurement to consider is calculating the cost per unit delivered. Simply add the direct fixed and variable costs required to perform a service, or to maintain a service over a measured amount of time and then subtract that amount from the revenue gained from the value stream during the same period.

For example, an on-site software installation could be itemized into the following categories:

- Travel, Hotel, and Meals for all involved on-site Process Owners.
- Total hours planning the install.
- Total hours away from the office implementing the install.

Summed up and subtracted from the price of implementing the contract, you'll get the cost per unit delivered.

For ongoing services direct fixed and variable costs such as operator hours, and equipment configuration time can be combined with indirect costs such as utilities to come up with a cost per daily unit of service.

It may be helpful for your department to obtain a standard burden rate from your finance department for some of these factors.

The Process Kata

As a Process Owner works his or her repeatable processes over and over, we call each iteration a Process Kata. The more you iterate a process, the better you get at it. Eventually, you can perform the process almost without thinking about it.

A Process Kata consists of the following steps:

1. Clarifying what the customer expects.
2. Starting at the beginning of each process with a Kata Card dedicated to that process.
3. If the process has a supplier process, collecting the incoming upstream Kata Card from the supplier and making sure it is complete.
4. Performing the process, as described in the Standard Operating Procedure, completing the Kata Card along the way.
5. As unexpected challenges arise, immediately consulting the recovery model for that process or the problematic asset in that process, and then executing the recovery steps if available, or performing a workaround, and then a root-cause analysis to determine the source of the problem. Once known, the problematic incident description, root cause, and recovery steps should be added to the process or asset recovery model to assist with future problem recoveries.

6. When an internal process is complete, record any tracked data on the Kata Card, and then hand off the completed Kata Card to the internal customer at the receiver process checkpoint, or the Master Chief if you are at the end of the value stream. It's helpful to ask the customer if he or she is satisfied with the results and if there are any additional improvement requests. It may be helpful to have a survey ready during this exchange.

The Coaching Kata

A Coaching Kata is performed when an existing process must be upgraded to achieve a more desirable result. A Coaching Kata consists of discussing the following questions:

1. What is the present condition compared to the desired situation?
2. What needs to happen to reach the desired condition?
3. What models exist, or who could we learn from that has already achieved similar results?
4. What is the next step?
5. How soon can learn from the results of the next step?

The target service level should be written in the Master Services List in the Configuration Management System and should also be specified on the associated process SLAM Chart.

It's important when making improvements to any system that you implement the improvements one at a time, or at least independently. If you improve several dependent factors at the same time you won't know which factor produced which effect. The total effect may even be negative, and you won't know which change caused the negative impact. For this reason, it's important to make only one dependent improvement at a time.

Solving Process Problems

Process problems can occur from all sorts of causes. Some examples include confused goals, insufficient system-wide understandings, flawed designs, sub-optimized goals and incentives, communication breakdowns, inadequate feedback, poor cooperation, and lack of accountability.

As problems occur—and they naturally will—the Process Owner adds an entry to that process Kata Card, to remind all future Process Owners to watch out for that problem going forward. This is real continuous improvement.

Process Owners become expert problem solvers. They learn to harness the power of teams and their extended resources through Kaizen activities. They identify and learn from others who have succeeded before them or have other relevant experience that may be applicable.

Eliminating Problems Using Kata Cards

At the core of continuous improvement is the Kata Card. As items are discovered that can cause problems if undetected, they are added to the process Kata Card and possibly the Standard Operating Procedure. From that time forth, the process should be protected from that known problem as the Process Owner uses the updated Kata Card. This is the philosophy of continuous improvement. Any problem can impact a process, but the same problem should never impact a process twice.

Effective Kata Cards have three sections: What to do before the event, what to do during the event, and what to do after the event.

The following figure is an example of a Kata Card. Notice it is organized by what to do before the event, during the event, and after the event. Notice also that this Kata Card encompasses two process steps, with signatures for each part. Be sure your Kata Cards have revision numbers or revision timestamps.

> This is the philosophy of **continuous improvement**. Any problem can impact a process, but the same problem should never impact a process twice.

It's also critical that Kata Cards get collected and archived. Therefore,

Kata Cards should get handed off to the receiver process at each checkpoint until the final Value Stream checkpoint, where the Master Chief receives the stack of Kata Cards and archives them.

Release Kata   Version 1.0

Software Release Version Number: _____

Build Kata Performed by: _____ Date/Time: _____

    Before Build:

    1. _ Software Checked In.

    Build:

    2. _ Build completed successfully. New Build Number: _____. Install package created.

    After Build:

    3. _ Test Folder "X:\TestPickup" erased/recreated and new install package copied to it.
    4. _ Email sent to tester notifying them of new install package available.

Test Kata Performed by: _____ Date/Time: _____

    Before Test:

    1. _ Install Destination folder erased/recreated.

    Test:

    2. _ Install package executed.
    3. _ Verify build matches Build Number above.
    4. _ Verify release notes match version number above.
    5. _ Verify help screen matches version number above.
    6. _ Test performed. Pass or Fail

    After Test:

    7. _ If Fail, return to Developer
    8. _ If Pass, execute Anti-virus check.
    9. _ If Pass, move Install Package to Release folder "Z:\Prod".
    10. _ If Pass, email team notification of success.

Change Control Approval Date: _____

Change Control Signatures:

Figure 32 - Example Kata Card

Improving Processes Through Critical Thinking

Critical thinking is an important skill to develop as a Process Owner. Critical thinking means that you have considered every possible angle to a situation, both popular and unpopular. It means you have gathered all conventional and alternative information and have weighed the costs and benefits. The opposite of critical thinking is biased thinking. Critical thinking is a life skill, and an essential skill in problem-solving.

While the list of cognitive biases is long and interesting, the most common biases that inhibit critical thinking are listed here:

1. Naivety Bias: Inexperience leads to poor decision making. Believing everything without checking for supporting or conflicting facts leads to errors in judgment.
2. Groupthink Bias: The tendency to do or believe something because many other people do.
3. Confirmation Bias: The tendency to search for or interpret information in a way that confirms your own preconceptions.
4. Extreme Aversion: The tendency to not pursue the best choice because it is less convenient than a more convenient lesser choice.
5. Planning Fallacy: The tendency to underestimate task-completion times.
6. Overconfidence Fallacy: The tendency to overstate the team's ability to perform.
7. Reactance Bias: The urge to do the opposite of what someone wants you to do out of a need to resist a perceived attempt by the other party to constrain your freedom of choice.
8. Status Quo Bias: The tendency for people to like things to stay relatively the same.
9. Anchoring Bias: The tendency to interpret a scenario based on past personal experience, which may not be the same for others.
10. First Impression Bias: Disproportional weight is given to the first information received.
11. Sunk-cost Bias: Bad ideas are considered still valid and worth further investment because a lot of money has been spent on them up until now.
12. Framing Bias: A situation is inaccurately explained second-hand, steering the listeners towards a pre-set conclusion.
13. Not Invented Here: The notion that we can build it better than anyone else. If we have it built somewhere else, it will be inferior.

The best defense against all these biases is awareness of reality. Therefore, measurement is a critical tool. When done correctly, measurement gets you closer to reality and steers you through all these biases so that decisions made are better over time. The better you are at making decisions, the more competitive you will become.

> The best defense against all of these biases is **awareness of reality**. This is why measurement is a critical tool. When done correctly, **measurement gets you closer to reality** and steers you through all of these biases so that your decisions are better over time.

Improving Process Through Nominal Group Technique

Nominal Group Technique (NGT) is a group brainstorming activity which is balanced to get feedback from everyone, even the quiet participants, and throttle the feedback from the louder participants.

In this model, the group splits up into smaller groups of 3-5 people, depending on the overall size of the combined group. A facilitator explains the problem or challenge at hand, and then the individuals in each group proceed to write down ideas individually on paper.

The participants within the groups then, one-by-one, disclose their ideas to each other in the first round of collaboration. The ideas are voted on and the ideas with the highest votes are then presented to the whole group in a second round of collaboration.

If the group was small to begin with then smaller groups breakouts are not necessary.

Identifying Problem Sources Through Root Cause Analysis

Root Cause Analysis is a popular problem-solving process that can be performed using several different techniques. The idea is to find out the real reason why a problem has occurred.

Often an immediate reason given for a problem might just be covering the real cause of the problem. At other times, there may be no obvious reason readily identifiable, so we must investigate. Following are some techniques that are useful in performing a root cause analysis.

Performing a Root Cause Analysis Using The 5 Why's Technique

One simple but effective technique for root cause analysis is called 5 Why's. Using this technique, the Process Owner repeatedly asks, "Why did this happen?" until the actual source of the problem is identified.

For example, when the famous Hubble Telescope was blasted up into space in 1990, its large mirror was found to have been shaped wrong, crippling the telescope's existing optics.

In this scenario, the Process Owner interacting with the team would ask questions in this order:

1. "Why were the optics crippled?"
   Answer: "Because the mirror was shaped wrong."
2. "Why was the mirror shaped wrong?"
   Answer "Because it was ground to the wrong specifications."
3. "Why were the specifications wrong?"
   Answer "Because we failed to account for the gravitational differences between Earth and space on the large glass mirror."
4. "Why did we fail to account for those difference?"
   Answer: "Because we could not think of a way to test the impacts of weightlessness on the glass mirror while it was still on the ground."

So, you see that within 5 questions, the root cause was uncovered. This was the actual explanation provided by the mirror manufacture. The optics in the Hubble telescope were corrected three years later in a subsequent shuttle flight.

In rare cases, it may take you even more than five questions to find the root cause, but you get the idea.

Performing A Root Cause Analysis Using an Ishikawa Diagram

Another root cause analysis tool that builds upon the 5 Why's technique is the Cause and Effect, or Ishikawa Diagram.

Process Step Categories

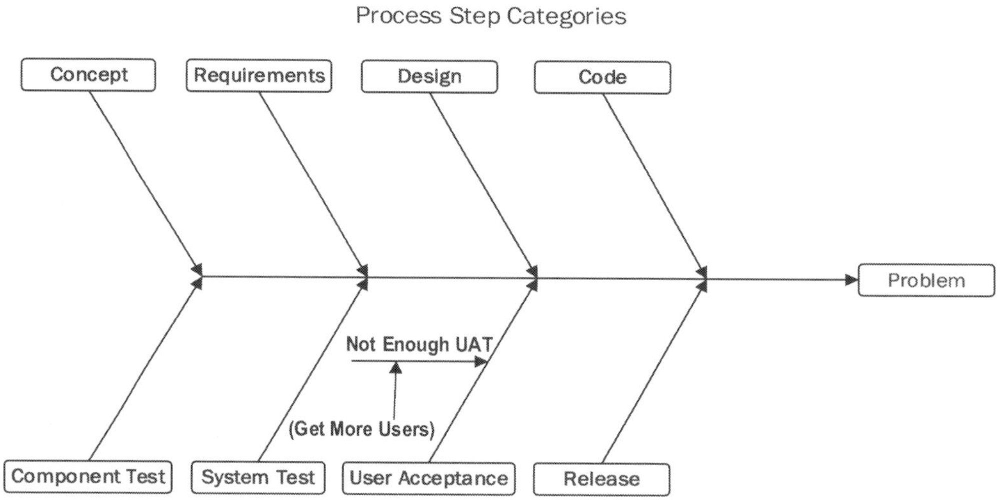

Figure 33 - Cause and Effect, or Ishikawa Diagram

Using a Cause and Effect diagram, combined with the 5 Why's technique, a group of people familiar with the problem can collectively brainstorm a root cause for a known problem.

It can be helpful to refer to or create a process flow model while identifying items of interest for the Cause-and-Effect Diagram. In addition, creating an Interrelationship Diagram is helpful while populating a Cause-and-Effect Diagram. With the benefit of a group discussion, several potential causes can be identified and prioritized in terms of what steps to take in what order to improve the system.

Afterwards, a Process Decision Program Chart (PDPC) can be created to address each specific problem.

Performing a Root Cause Analysis Using a Focus Group

Focus Groups are meetings called by a Process Owner who invites others to the meeting who have experienced a problem or talked with customers who have experienced a problem.

The purpose of a Focus Group Meeting is to understand a problem better by asking questions to those closest to the problem. As the PRO facilitates, the invited guests are asked to describe what they experienced. The participants search for patterns that may provide clues

A3 No, Name, Date	Process Owner	Team Members
1. Problem Statement		5. Corrective Action(s)
2. Current Condition		6. Preventative Action(s)
3. Target Condition		7. Results Summary
4. Root Cause Analysis		8. Follow-up Action(s)

Figure 34 - A3 Problem Solving Template

to eventually identify what went wrong.

The meeting is for information gathering purposes, and not specifically to solve the problem while in the meeting. One of the other techniques listed in this section could be conducted after a Focus Group, such as the 5 whys, or an After-Action Review, to get to the root cause. You don't really want to expend customers' time on technical troubleshooting if it can be avoided.

Performing a Root Cause Analysis using the A3 Problem Solving Approach

An elaborate technique for identifying and solving a problem is the A3 Problem Solving Approach. With this approach, the team uses a single sheet of A3 sized paper (11" x 17") to show all the details about a problem, including remediation plans.

Various templates exist on the internet containing a simple logical flow of information gathering and analysis steps including a problem statement, current conditions, target condition, root cause analysis, corrective actions, preventative actions, results summary, and follow-up actions.

Most of the fields in the example template on the previous page are self-explanatory. Remember that Corrective Action is what needs to change within the target system, while Preventative Action is what changes can be made to additional systems to achieve similar improvements. The A3 format is an excellent template for Kaizen Projects.

Improving Customer Engagement using After Action Reviews

An After-Action Review (AAR) is a retrospective meeting held by the team directly after an engagement with a customer. They can also be called for other purposes as well, such as Root Cause Analysis, or after a major troubling event.

In an After-Action Review, the team shares observations, questions, and lessons learned about what went well, and what could have gone better. It is targeted at a specific event and should be conducted directly after the event when the team's observations are fresh in their memories.

Be sure any lessons learned are recorded in the Process or Asset Controls.

Improving Customer Engagement using a Facilitated Workshop

A Facilitated Workshop can be called for many purposes. These meetings are excellent tools for brainstorming, knowledge transfers, walkthroughs, requirements elicitation, and many other purposes.

A Process Owner calls a facilitated workshop and invites customers and technical engineers, if needed, to the meeting. These meetings tend to be longer and some can span multiple days.

Risk Management using a Failure Mode and Effects Analysis (FMEA)

A thorough technique used in military-grade development and operations planning is a tool called a Failure Modes and Effects Analysis (FMEA).

In this exercise, a spreadsheet is created listing every component of a system or process, and the next column over lists every potential way it could fail. Then, the impacts of each failure are listed in the next column, and each impact is assigned a probability and severity rating.

Detection techniques and mitigation steps are listed, and the analysis becomes a reference sheet for designers of the system and Process Owners going forward.

After an FMEA is conducted, a Process Design Program Chart (PDPC) can be created, identifying risk meditation strategies for specific high-risk areas. See Chapter 23 – Data Collection and Analysis Tools for a description and an example of a PDPC.

The Future Domain and the Process Owner

The Process Owner is the closest to the work and is responsible for creating and maintaining the SOP's if they don't already exist. In my experience, SOP's rarely get read by employees beyond required training.

The practical purpose for an SOP is to ensure consistency. They should be read by new employees during initial employee onboarding and then be reviewed once or twice a year and brought up to the current standard, as needed.

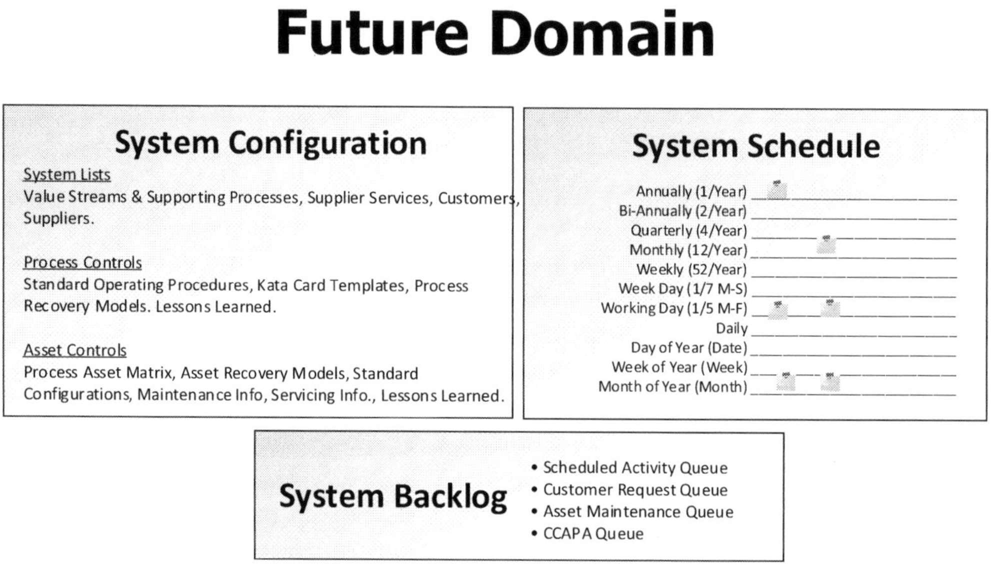

Future Domain

System Configuration	System Schedule
System Lists Value Streams & Supporting Processes, Supplier Services, Customers, Suppliers. **Process Controls** Standard Operating Procedures, Kata Card Templates, Process Recovery Models. Lessons Learned. **Asset Controls** Process Asset Matrix, Asset Recovery Models, Standard Configurations, Maintenance Info, Servicing Info., Lessons Learned.	Annually (1/Year) _____ Bi-Annually (2/Year) _____ Quarterly (4/Year) _____ Monthly (12/Year) _____ Weekly (52/Year) _____ Week Day (1/7 M-S) _____ Working Day (1/5 M-F) _____ Daily _____ Day of Year (Date) _____ Week of Year (Week) _____ Month of Year (Month) _____

System Backlog
- Scheduled Activity Queue
- Customer Request Queue
- Asset Maintenance Queue
- CCAPA Queue

Figure 35 - The Future Domain

An SOP should always represent the best-known way to execute a process. The process Kata Card should be filled out with every Stable Process Kata. The Kata Cards should be updated regularly as any new failures are experienced and root causes are identified during normal work processes. All training should be recorded in the historical training log by the Master Chief.

For groups just adopting Stable, the Quality Planning phase is where all of this gets created. Here the PRO's create their SOP's and Kata Cards using the Standardize-Do-Check-Act model, based on the best known-method for performing each process.

A list of primary customers for their processes and their contact information should be made by the Process Owner and kept in a Customer List in the Configuration Management System. Initially, the PRO should meet with all or a representative sampling of the customers and establish the relationship and confirm expectations.

An agreement should be made on how the PRO will get regular customer feedback and get any new feature requests. Depending on circumstances this can be a regular survey, phone call, visit, golf game, lunch, hallway conversation, or whatever seems appropriate. Whatever form it takes, what's important is that it is done regularly. A good rule of thumb is to confirm expectations have been met at every point of contact with the customer.

Deming would help employees connect with their customers regularly. He would shut down entire assembly lines and have all the workers listen in to a conference call with random customers.

He would ask the customers how the company's products and services helped the customers and how they could be improved. This is important because it kept the employees emotionally bonded with the customers, which improved what Deming called "Pride in workmanship."

It's healthy for the PRO's to meet the people they are helping, and it helps the customers feel emotionally bonded to your organization. This a critical success factor in raising your Net Promoter Score among you customers.

Customer Satisfaction and Net Promoter Score

It turns out that Customer Satisfaction ratings are important, but not the most important customer metric to a business. Surprisingly, how well a customer has emotionally bonded to your company is a better indicator of future business than only having good customer satisfaction scores.

Customers might give you high satisfaction ratings, but not necessarily refer anyone to your business. It requires an emotional connection for them to take the initiative and refer their associates to you.

The probability for a customer to recommend your services to someone else is called a Net Promoter Score (NPS). The Net Promoter Score can be calculated by asking your customers in an anonymous survey, on a scale of 0 to 10, how likely they are to recommend your company to another colleague. Depending on how they respond, they fall into several categories:

- 0-6 detractors
- 7-8 passives
- 9-10 promoters

Then, to calculate the score, use this formula: NPS = % promoters - % detractors. The larger the positive number, the better.

So how do you get your customers to emotionally bond with you? Emotional bonding happens when customers are addressed and respected as the individuals that they are. Surveys and emails should be personalized. Feedback from them should be answered personally, and quickly.

Your customers should get a consistent message that you are always aware of them and care about their needs and their successes. Teddy Roosevelt was famous for saying, *"People don't care how much you know until they know how much you care."* Ask your customers about their individual needs and then discuss how your services could better address them, rather than talking only to your customers about your services.

> Ask your customers about their **individual needs** and then discuss how your service could better address them, rather than talking only about your services to your customers.

The best brands have figured out that they are really selling feelings, not features. Think about Nike and you can imagine yourself as someone who is fit and exercises every day. Think about any of the prominent insurance companies and you'll get a feeling of protection. Think about driving a Volkswagen and you'll suddenly be thinking of yourself as a young urban professional with a lot of friends.

There is an entire industry devoted to helping create an emotional connection with your customers. You will grow your customer base by learning from it.

The PRO is also responsible for compiling a Process Asset Matrix linking the Process Owners Processes to required Assets, and also an Assets List containing all the primary assets responsible for performing the PRO's repeatable processes. For each asset, an asset recovery model and maintenance plan should be created, if needed.

The Present Domain and the Process Owner

The Master Chief is responsible for populating items from the System Schedule and System Backlog into the "Objectives" column of the Kanban board during the Cycle Planning Meeting. As Process Owners complete existing work items and are freed up for more work, they can grab items from the top of the "Objectives" column. During the Master Cycle, they will naturally have additional process work items emerge that they need to work on. They should note these items on the Kanban so that they can be accounted for at the end of the cycle.

As Process Owners finish existing items, they can grab the next item they are qualified to perform from the "Objectives" column.

As they work the items, they move them from the "Selected Tasks To Do" column to the "In Process" column and finally to the "Completed" column. Some groups have a "On Hold" column to deal with larger objectives that get held up or require senior management to expedite.

Present Domain

Team
- Daily Kaizen Standup
- S5 Housekeeping
- Adhoc Kaizen Teams
- Process Kata & Kata Cards
- Training/ReTraining

Committees
- Change Control Committee
- CCAPA Committee

Environment
- Alert System
- Production Change Log
- Suggestion Box

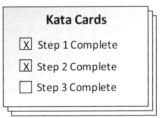

Figure 36 - The Present Domain

Items should be time stamped as they move from the "Selected Tasks To Do" to the "In Process" and finally the "Done" columns. Kanban Boards allow us to measure how many of each type of work items that were completed during the Master Cycle, and the average time required to complete each type of task. This provides historical data that can be used for forecasting future performance for customers.

The team should be encouraged to complete what they are working on before turning their attention to something new. Task switching forces us to move our mental assets around in our minds and is a form of inventory, motion, and transport waste.

When problems occur while working a process, the Process Owner should identify the process or asset containing the problem and go directly to its respective Process or Asset Recovery Model (PRM, or ARM Sheet). This is a list stored in the Process Controls or Asset Controls portion of the Configuration Management System and contains descriptions, root causes, and solutions to past problems associated with the process or asset. When new problems occur, their root causes, and steps to fix them get added to the PRM or ARM sheet. By consulting these Recovery Models, repeated problems get resolved quickly.

The Past Domain and the Process Owner

The Process Owner is responsible for communicating the status of all the

Past Domain

Master Cycle Totals
- Cumulative System Perf. (CSP) Chart
- Supplier Services SLAM Chart
- Cycle Totals
- Customer Satisfaction
- Customer Net Promoter Score
- SLAM Chart A, B, etc.

Historical Logs
- Production Change Log
- Critical Decision Log
- Kata Card Historical Repository
- CSP Repository
- Training Log

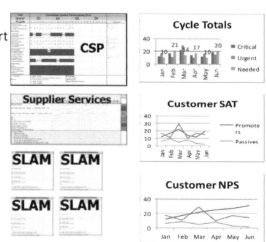

Figure 37 - The Past Domain

processes they are responsible for, during, and especially at the end of each Master Cycle.

This is done through measurement. Measurement is the essence of accountability and helps us understand what is going on in our environment.

The process measures described previously in this chapter get recorded and displayed visually on the Performance Console. This type of visual management is critical to achieving Operational Excellence. We want to be transparent about the health of our processes, customer relationships, service levels, and team morale so that our leaders whom we report to are free to spend their time growing the business and not wondering about our areas. We are trying to express this information so clearly that a visitor walking through our area could see our Performance Console and know how our department is doing.

Two Types of Measurement

There are two types of measurements that are important. One is measuring how close the results were to the target. The other is measuring how close the target is to what the customers want.

For example, the first type of measuring is detecting how close we can get the room temperature to 72 degrees if that was our target. The second is measuring to see if 72 is the most desired temperature for the people in the room. You want to see how accurate you were to the target, but you also want to know if the target is currently accurate.

In summary, we want to use data to determine if we are reaching our goals, and what the current success criteria is for those goals.

Visual Communication

Visual aids are most helpful. It's always better to use graphics and as few words as possible, but enough to be clear about what the graphics indicate.

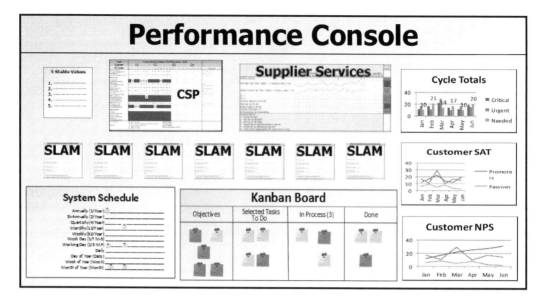

Figure 38 - The Performance Console

All these graphs and charts are displayed on the Performance Console, which should look like the previous figure.

Graphs and charts are excellent tools for showing performance over time. The final chapter in this book is called "Data Collection and Analysis Tools" and details a set of tools for collecting, analyzing, and reporting in various ways.

While creating your charts be sure that your graphics are not too distracting. Every single pixel competes with your message, so when reporting with graphics be sure to favor simple graphics over fancy graphics.

As a professional, every time you report on status you should be interested in two factors. One is the current status, the other is the trend. The trend is almost as important as the current status. Is the reported performance improving or degrading? You'll want to report on this, as well.

Service Level Attainment Monitors, or SLAM Charts make excellent reporting tools for a group of processes.

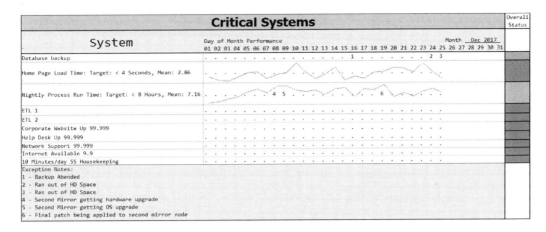

Figure 39 - Service Level Attainment Monitor (SLAM) Chart

The number of tasks completed can be categorized by type or urgency and listed over time in a bar graph, or a line chart.

In addition, any of the quality charts mentioned in the last chapter of this book can be used to report on specific information important to the organization.

Master Cycle Review Meeting

Status Charts are viewed with internal customers and senior management at the Master Cycle Review Meeting, which occurs at the end of each Master Cycle. During each Cycle Review Meeting, each team member reports on the service level performance of their processes and explains any outstanding challenges remaining. Customer Satisfaction results are also presented at the meeting, by each Process Owner.

CCAPA and Change Control

The CCAPA List status is presented by the Master Chief, as well as any outstanding Change Control items that need to come before the group if change control is handled at this meeting.

Kaizen Project Initiatives

Optionally, individuals can be challenged at each Master Cycle Retrospective to identify a Kaizen Project to improve something in one of their processes, or in their workplace. These should be small improvement efforts that take from a few minutes to few days of effort. Then, during the Master Cycle Review Meeting they should be tasked with reporting on the improved objective and explaining the benefit to the organization.

Master Cycle Retrospective Meeting

The Master Cycle Retrospective Meeting happens just after the Master Cycle Review Meeting. This meeting should last no longer than one hour. Two hours, perhaps, if the team engages in a Kaizen Project. Only the team of Process Owners and, optionally, the Master Chief, will attend. No customers should be present.

The team discusses these three questions:

1. What went well this past cycle?
2. What could have gone better this past cycle?
3. What still feeds the hidden factory?

These questions can lead to discussions and ideas that trigger new Kaizen Projects.

Chapter 13 - The Master Chief

The Master Chief, or MC, is responsible for the overall success of the value system. MCs are also responsible for ensuring that the qualitative work efforts expended within the environment get converted into quantitative information to be reported on at the end of each Master Cycle and then eventually into quarterly trend reports. Simply put, they are responsible to ensure that the Stable Framework™ is understood, implemented, and executed properly within their environments.

The MCs are the stewards of the value system. This means they are responsible for organizing all the value streams and then producing accountability among the Process Owners to achieve the desired results for each value proposition within their value streams. In addition, MCs are responsible for any necessary coordination between interactive value streams.

The key to producing accountability among the Process Owners is by creating a sense of ownership through involvement. The Master Chief must ensure the Process Owners are involved in the systemization, customer engagement, measurements, and improvement efforts of their systems. If the Master Chiefs cannot create ownership, they will not be effective in generating accountability.

The Master Chief must be flexible enough to stay informed and help assist anywhere when needed. This is a challenge and requires a lot of responsibility, competence, and connecting with people. Coaching, mentoring, and communication are the primary tools of the Master Chief. Plan on devoting time to become skilled in these areas.

Since Stable is applicable anywhere there is a team, there may be one Master Chief, or there may be several Master Chiefs in a larger functional unit working in coordination with each other under a Senior Master Chief. A Master Chief presides over an empowered team of Process Owners, with each Process Owner separately being responsible for the execution and accuracy of the repeatable processes they own. The Master Chief is responsible for facilitating the coordination of work between these processes.

Master Chiefs are responsible for the success of the entire system, and they achieve this by informing, equipping, empowering, coaching, mentoring, and holding their Process Owners accountable for the performance of their processes. All of this requires leadership.

Although the individual Process Owners are responsible for collecting actionable qualitative information about how their process deliverables can better meet their customers' needs, the Master Chief is responsible for collecting quantitative customer satisfaction, and Net Promoter Score metrics. If they don't collect it them self, they need to be sure Marketing, or the Service Desk, or someone other than the Process Owners are collecting it.

Customer satisfaction metrics should never be collected by the same people responsible for the scores.

Leadership

Leadership is one of the most written about topics in history. Like John C. Maxwell says, "Everything rises and falls on leadership." Leadership is influence. Therefore, to become a leader you must become a person of influence. To become a person of influence, you must become a person of value to those whom you report to, and to your Process Owners.

Leaders aren't leaders unless they have followers. Let's discuss the dynamics of leading people.

The Three Concerns of Team Leadership

As a Master Chief, you are ultimately responsible for the performance of the team. To succeed, you must concern yourself with three areas:

1. Team production results
2. Team cohesion
3. The individual needs of the people on your team

Production Velocity vs. Individual Needs

In 1964 Robert Blake and Jane Mouton developed their Managerial Grid model to articulate approaches for managing individual needs versus team needs.

They organized their grid into two axes: Concern for People, and Concern or Production. Their model contains five approaches to balancing these concerns. Each approach is explained below.

Managerial Grid

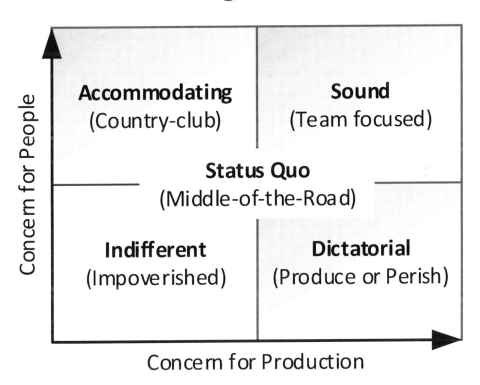

Figure 40 - The Blake-Mouton Managerial Grid

<u>Indifferent</u> leaders tend to hide and evade from the team as much as possible. By protecting themselves from getting involved and potentially being part of mistakes, they think they are preserving their seniority and job security. Innovation and conviction suffer in these environments.

<u>Accommodating</u> leaders demonstrate a high concern for people at the expense of production. These leaders pay close attention to security and comfort of their employees, but generally do not lead strongly productive environments.

Dictatorial leaders find employee needs mostly unimportant, and lead transaction-based environments. Rewards and penalties are emphasized, and employees are expected to do their jobs with minimal personal conflicts. This can be an effective balance during a crisis or in highly competitive production environments.

Status Quo leaders attempt to balance employee needs with production goals, making compromises in each area as needed. This is a reactive mentality and minimal leadership is devoted to making up the difference.

Sound leaders are transformational. When conflicts arise between concerns for people and production, sound leaders innovate to find ways to cater to employees while keeping commitments and meeting goals along the way. Sound leadership requires ingenuity, creativity, and innovation.

Production Velocity vs. Team Spirit

Kurt Lewin, the founder of early social psychology who coined the term "The whole is greater than the sum of its parts," helped us understand that within any team there needs to be two types of activities--those needed to achieve the team's objectives and those needed to develop and maintain team spirit.

Dr. Eddie O'Connor expounded on these ideas by identifying four components needed to achieve excellent team spirit: Cohesion, Cooperation, Role Relationships, and Leadership.

Cohesion is how willing a team is to take care of each other and work together particularly during adversity. Bonding with a group is satisfying and meets several basic human needs. Cohesion can be based on different things and usually evolves over time, making teambuilding activities an ongoing requirement for long term teams.

Cooperation requires the team member to emphasis the team above him or herself, which is essential for effective teamwork. To make this easier, it helps to have a common set of team values and goals. We'll discuss creating a team charter in an upcoming section about creating a team.

Role Relationships are important to establish and clarify. Ambiguity is the enemy of progress and will drain physical and mental effort in team members as they feel a sense of urgency, but don't know what to do next. This leads to confusion and then a lack of cohesion resulting in reduced performance. Be sure to clearly define roles and decision-making power as you interact with your teams.

Leadership styles have an immense impact on team spirit and velocity. The next section is an overview of different approaches to leading teams. The differences are important to understand because different leadership styles affect the way team members feel about themselves. Be sure your style and words build confidence. Catch members doing something right and compliment them for it. Celebrate achievements and make it apparent often that you believe in your people.

Leadership Styles

In the past, leadership was thought of as an attribute that only some people were born with and the rest don't have. Today, the thinking has changed, and leadership has become more of a science. Here is a brief description of the more common leadership styles identified today:

Transformational Leadership is thought of as the most effective style to use. Transformational leaders inspire their team members because they expect the best from everyone, and hold everyone including themselves accountable for their actions. They demonstrate high integrity, emotional intelligence, a shared vision of the future, and they communicate well. They are self-aware, authentic, empathetic, and humble. They set clear goals and are good at resolving conflicts.

Transactional Leadership is influence by mandate. When an employee accepts a job, they agree to follow a leader. Transactional leaders enjoy influence due to position or title, and have the right to apply penalties if work standards are not met. This leadership style is ideal for employees seeking performance-based rewards within a structured environment. The high structure and rigid expectations, however, can curb creativity and innovation.

Autocratic Leadership is like transformational leadership except that the focus is on the approach the leader wants to take to achieve the team goals, instead of the approach the team wants to take. Although sometimes appropriate for short term tasks, this approach compromises team buy-in and is demoralizing during long term projects. This is sometimes called Charismatic Leadership.

<u>Democratic Leadership</u> is like autocratic leadership except that the team is solicited for ideas and feedback before a final decision is made by the leader.

<u>Bureaucratic Leadership</u> is present in high-risk, volatile environments. Work performed with toxic waste, expensive manufacturing, or the handling of large sums of money requires checklists and checkpoints to protect against unnecessary mistakes or misfortune. This type of leadership must be applied carefully so as not to restrict innovation and creativity where needed.

<u>Laissez-Faire Leadership</u> is where a leader offers little or no guidance to group members and leaves decision making completely up to the group. This approach works well with a team of highly-qualified experts, but can also lead to low productivity and high-maintenance team members.

<u>Servant Leadership</u> is a facilitative-based leadership style where the leader clarifies objectives, constraints, roles, and expectations, and then challenges the team to come up with the plan to achieve the objectives. Then, the servant leader uses his or her authority to empower the team and remove roadblocks from their path.

Based on the situation, some styles are more appropriate than others. Like having a set of golf clubs, a leader will be most effective if they learn to use the right club at the right time.

The Stable Framework™ Leadership Model

No matter which leadership style is applied, Master Chiefs must develop the qualities shown in the Stable Leadership Model below, and then use them to succeed within their teams.

Warren Buffet said, "A leader is someone who can get things done through other people." To understand leadership better, it's useful to understand why people follow leaders.

Gallop Research interviewed 10,000 followers across a broad spectrum of society and asked them what they experienced from the most influential leaders in their past. The four most common words were: Stability, Hope,

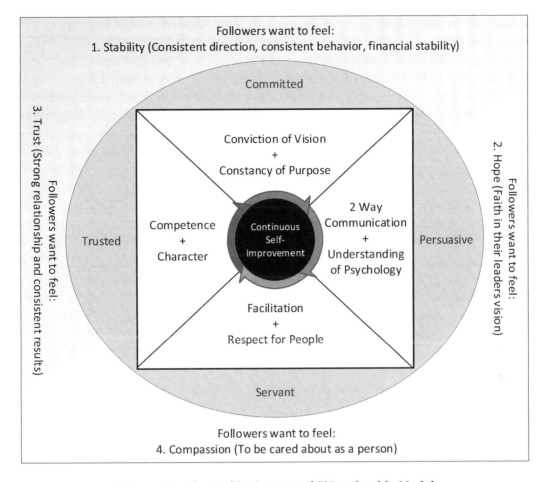

Followers want to feel:
1. Stability (Consistent direction, consistent behavior, financial stability)

Committed

Conviction of Vision
+
Constancy of Purpose

Competence
+
Character

Continuous
Self-
Improvement

2 Way
Communication
+
Understanding
of Psychology

Trusted

Persuasive

Facilitation
+
Respect for People

Servant

3. Trust (Strong relationship and consistent results)

Followers want to feel:

2. Hope (Faith in their leaders vision)

Followers want to feel:

Followers want to feel:
4. Compassion (To be cared about as a person)

Figure 41 - The Stable Framework™ Leadership Model

Trust, and Compassion. As a Master Chief, remember that your behavior, focus, and communication must create these four feelings within your people. The Stable Leadership Model will help you with this.

Committed

The first step in leadership is being *committed* to a purpose. Leadership is about moving people from one state to another, and it's easiest to do that if your people understand why a change is needed. A leader must have a vision of the future and believe that the future is in the group's best interest and be able to express why it's in the group's best interest. Deming emphasized the importance of connecting workers with the customers their labors are helping. He said it is critical for them to understand the benefits their labors are contributing to the lives of the customers. This instills pride of workmanship and motivation in the teams.

Once the leader establishes the vision or purpose, the leader must stay constantly focused on it. The leader's job is to have a vision, be sure the team clearly understands that vision, and help them be enthusiastic about it.

Deming talked about the importance of maintaining a constancy of purpose with your efforts towards your vision. Relentless focus while working towards that vision is how great things come about.

As a Master Chief, be sure your team understands your conviction for consistently improving customer understanding, customer experience, and system improvements using Stable principles. Be sure they understand it is a never-ending cycle. Your conviction will catch on as you transfer your energy into the team.

Trusted

A leader must be trusted. Your people must feel safe in your presence and in the culture that you create so that they can be transparent and emotionally vulnerable with you and with each other. Emotional vulnerability is one of the primary elements of high-performance teams because it connects people to each other and reduces the noisy emotion and politics that slow down group trust and progress.

Trust is generated by a leader who demonstrates competence and character over time. Competence is the ability to read a situation accurately and apply the correct agenda, while Character is the assurance that a leader will consistently do the right thing, whether that team member is around or not. A history of observations by your team members that you are competent and have character will produce trust by your team in you.

As a Master Chief, your past personal experience in your business domain and skills will add to your competence. Be sure you are a person of character. The combination of the two will enable people to trust you.

As a trusted leader your Process Owners will feel safe to work together in the culture you create and support.

Persuasive

A leader must have followers, or they are not a leader. In order to have follows a leader must be *persuasive*. Persuasion is about communication and psychology. A leader must be able to communicate his or her vision and understand how people think and reason in order to put the vision message in the right context.

A good leader can express his or her message so that it calls people to action. This is done by validating the challenges and pain points of the leader's team or customers and then stating an improved future vision that the leader believes is attainable. Depending on the situation, the leader can then provide a solution (coaching) or challenge the team to create a solution (mentoring) that will bring the team from the current state to the desired state. This is a basic model for sales and is the basic formula for persuasion.

As a Master Chief helping your team succeed through Stable principles, you'll often need to persuade your people to stay on track. You'll need to help them understand the value of spending a little more time with their customers which will emotionally connect their customers to their work. You'll need to persuade them to understand that a little more time spent doing rudimentary quality tasks such as completing checklists is much better than spending unscheduled time putting out fires and reworking inaccurate efforts for a second or third time.

Servant

An emphasized leadership style in the Stable Framework™ leadership model is Servant Leadership. Servant Leadership is a concept put forth by Robert K. Greenleaf in a book by the same name written in 1977. In his book, Greenleaf suggests that the best leaders empower their peers and direct reports and help them become leaders themselves. They give them objectives and constraints but challenge them to solve their own problems. Servant Leaders use their authority to remove roadblocks from their teams paths instead of commanding them in every matter.

Servant Leadership requires respect for people and facilitation skills. As a Master Chief, you must believe and trust that your people can generate excellent ideas and solutions to problems. Avoid the impulse to provide the answers, and instead use curiosity and involvement as tools to get your people bought in. Get them to come up with options by asking questions like:

- What's the best way to get started?
- What are the major steps along the way?
- What kinds of risks might we encounter if we did that?
- What other options can you think of?
- Who else could you ask?
- How would we know if it's the best decision?
- Is there an option that could get us there faster?
- Could we achieve the same result some other way without spending any money?
- Who wants to take the first step?

Jack Welch, the retired CEO of General Electric, wrote about this technique in his book, *Winning*. He called it the "Dumbest guy in the room." technique. He said you ask all the questions and soon your team has come up with the entire plan. If you do it right when you leave, they will say, "What did we need our leader for? We came up with the whole plan by ourselves."

Facilitation is an important part of servant leadership. A facilitator deals with the meta aspects of interpersonal behavior, initiates conversations and meetings, creates appropriate space for dialogue and discussion, and uses gatekeeping skills to encourage quieter people to share more while throttling back the continuous energy of the more outspoken participants.

A facilitator also works behind the scenes to prepare for these events, and to remove roadblocks from his or her team's path when possible. As a Master Chief, you will constantly be facilitating for your teams.

Continuous Self-Improvement

At the heart of a successful Master Chief is continuous self-improvement. A Master Chief is always learning, and always favoring improved ways to deliver value over time. Master Chiefs must also encourage continuous improvement in their environment and within their people. Master Chiefs will be unable to do this well unless they are always improving themselves along the way.

Deming talked about personal learning being a foundation of leadership in a group setting. Constantly share with your team what you are studying and learning about. Encourage book clubs and brown-bag lunch sessions where your Process Owners are instructing each other on better ways to live life. Make the workplace an asset in your team members' lives.

In Jim Collins book, *Good to Great*, his team of researchers analyzed leaders of top performing companies and discovered that most of these effective leaders don't profile as magnanimous giants of industry. In fact, quite the opposite is true. Most of the effective leaders his group examined were humble, down-to-earth performers, who had constructed effective teams. This is a worth noting.

Adolescent leaders try to do everything alone and not tap into their teams. In some cases, they even wastefully compete with their own teams. As leaders mature, they learn to harness the power of their teams by empowering the teams to make the important decisions and showing them that they trust their team's judgment.

Deming is famous for his fourteen points of leadership, which he writes about in his book, *Out of the Crisis*. Every Master Chief and Process Owner should familiarize themselves with these points. They make an excellent set of topics for a book club or regular discussion.

> Adolescent leaders try to do everything alone and not tap into their teams. In some cases, they even compete with their own teams. As leaders mature, they learn to harness the power of their teams by empowering them to make the important decisions and showing them that they trust their team's judgment.

Peter Druker, the famous management educator and economist, described the presence a leader needs to have. He said, first, that leadership is not a position, but rather it is a means to get some future objectives accomplished. Leaders should see leadership as a responsibility, and not rank and privilege. When things go wrong—and they will at times—the leader must own the situation and blame them self. Be prepared for things to go wrong occasionally—and be prepared to own the failure and blame yourself, not your team. Failure can be an awesome learning process, but if the professional will not own the failure, they will have learned nothing and, given the same scenario in the future, will repeat the same mistakes.

In addition, if you own the mistakes of your team, you will experience great joy and exhilaration in their successes. Be sure to always give credit for successes to your team. This protects you from competing with them as their leader.

A leader must be visible. They must be front and center. They must own the airspace. If they don't own the airspace, it will be filled with other things from other sources, which will confuse the team.

A leader must be right. Leaders that are wrong quickly erode the patience of the team. Some of the greatest leaders in history had no charisma but were right. Look at Albert Einstein during World War II who got the attention of President Roosevelt to warn him that science had advanced enough that the Axis powers could be working on the development of the first Atomic Bomb. He was right.

A leader must also understand that the domain size of his or her own areas of connection and influence exceeds the domain size of any of their direct reports. While direct reports have their own workplace environments, the leader's environment is always larger. This gives the leader access to people, information, and resources the direct reports don't have. Respect this difference and own it. This additional reach empowers you to help your team succeed, if you use that reach wisely.

In addition, be aware that as people we are all equals, but as roles we are not. Inside the building, your word has greater application then theirs, but outside in the parking lot you are both equal human beings.

Find out about their career goals and life goals and help them any way you can. Remember that you are working with someone's sister, or someone's dad, or someone's husband. Be sure your people know that you know they have families outside of the workplace that need them, too. It's okay to learn about people's personal lives and provide support. Help them understand you want them to succeed in life just as much as in the workplace.

John C. Maxwell talks about how people will follow a leader because of his or her title for a short time, but they are really watching the leader to determine if the leader wants them to succeed. If they believe their leader wants them to succeed, they will follow that leader for many reasons.

Personal Leadership

Four elements are fundamental to career success:

1. Communication and networking skills
2. A strong work ethic
3. Good decision-making skills
4. The ability to deliver results with your team

The challenge is, if you focus on any one of these too much, the others will suffer. As a leader, you must find quiet time to reflect on your goals in these areas and be sure you are not neglecting any of them.

Periodically ask yourself if you have any working relationship that need attention. Are there any skills that you need to develop or polish up on? Is there a problem developing that you can address now? Is your current performance measurement system addressing all the questions that your leaders ask? Stay focused on these questions as you work with the PRO's coordinating the Value Stream results in your area.

Building the Right Culture

Culture is defined as the anticipated reactions between two or more people, and it varies from place to place. In America, for example, when somebody says, "What's up?" the only proper reply is the exact same expression, whereas in another country the question would be confusing.

Just as international cultures differ, cultures exist and differ from one work environment to the next. Culture is a combination of group norms, group history, elements in the environment, and the personalities that each team member possesses in the group.

In a workplace environment, culture is often categorized along two continuums: Independent vs. Team Focused, and Flexible vs. Rigid. Your organization can be anywhere within these two continuums.

Independent cultures tend to be more competitive whereas team focused environments tend to be more supportive. Rigid cultures tend to have more inertia and be less tolerant of failure, whereas flexible environments allow for more innovation, creativity, and new ideas to flourish.

Independent cultures tend to produce more work over time, whereas team focused cultures tend to make safer decisions.

The beauty of the Stable Framework™ is that it offers stability in the right places, allowing for innovation and flexibility towards improvement over time.

> Independent cultures tend to produce more work over time, whereas team focused cultures tend to make safer decisions.

It also offers the team a model for working independently yet harnessing the best minds in the group for Kaizen Blitz problem solving, and Kaizen Improvement Projects. In other words, it provides mechanisms that support all four extremes.

Changing a Workplace Culture

You may be starting with a group that is not used to teamwork, or not used to flexibility. One of my clients was appointed senior project manager over a team of experienced project managers. She was surprised that her new team wanted assistance making decisions about every detail in their projects. My client reported to me that the team members acted confused when asked to make a decision. I asked my client how the team described the previous manager and I was told "an extreme micromanager." I pointed out to her that the team was used to that culture and was not yet ready to be given the flexibility to make their own decisions but would require her leadership to shift into that paradigm. She was experiencing a culture clash.

Changing culture is historically challenging. Many corporate transformations failed after the employees were trained properly, and after the new tools were put in place, simply because the culture had not been addressed.

Transformations usually require a change in culture to be successful. Therefore, as a Master Chief you must know how to do this. Fortunately, there is a powerful set of tools for changing the culture within an organization.

First and most powerful is the creation of team memories. Memories manufactured over time, with the same group, create a sense of a team cohesion. It doesn't matter what the memories are, it's just important that they are frequent, and hopefully positive.

Eat together. Have a piñata activity after each significant victory. Get out of the building and have a movie night for team members and their spouses. People will tell you things outside of the building that they will never say inside of it.

Next, talk about desired outcomes. Culture is also shaped by what people talk about. Own the airspace. State the obvious. Become a broken record by repeating the desired outcomes. Say it so many times that your team members start to expect to hear it from you.

Pick your team members carefully. Jim Collins famously talked about having the right people on the bus. Look for team members who champion where you want to go and give them prominence. Draw attention to their victories implementing the new ideas.

Remind the teams that you are going through a transformation and it's healthy to talk about what's different now, and why it was needed. Encourage them to rely on each other for help.

Finally, in some cases, you may have to move people and furniture to help support the team.

Establishing team spirit and changing culture is empowering. These principles work not just in the workplace, but in family dynamics, neighborhood relations, and beyond.

Creating a Team

A great beginning with any team is a
Team Charter creation ceremony. Make
this a memorable event. Call a meeting
specifically for drafting a team charter.
A team charter consists of a vision,
mission, values, and goals statements.

TOGETHER
EVERYONE
ACHIEVES
MORE

Of course, the objectives of the team are pre-set by whoever organized
the team, but the team is free to decide how they are going to operate
along the way.

An Identity Statement is an expression of who the team members are.
What do they identify with? Are they the best developers of their
codebase in their area? Are the video-gamers turned coders? Are they
athletes with programming jobs? Who do they identity with?

A Vision Statement is a description of a future state the team wants to
bring to pass. For example, the team might agree that they want their
sponsors to think they are the best team they've ever had, or that they
are all great friends a year from now, or that they will all be masters of
certain aspects of their trades by next year.

A Mission Statement is a description of how the team will accomplish
their objectives and their vision. For example, they might say "We will
continually learn about our customers, innovate, and improve our
processes to make our services better, and we will take care of each other
along the way to make our workplace relationships better over time."

Our Team Charter

- Identity Statement
- Vision Statement
- Mission Statement
- Team Values
- Team Goals (Short Term & Long Term)

A Value Statement is a short list of prominent values the team agrees will govern their behavior. Values like respect for people and punctuality are good examples of team values.

Finally, a Goal Statement might be a combination of long-term and short-term goals. These might be to raise customer satisfaction ratings monthly or to have all the critical processes listed on the performance console.

Have everybody sign and date the team charter. Frame it and hang it on the wall in a public place. As new employees join the team, take the charter down and have the team re-evaluate the charter together with the new employee. It will be a great memory supporting the new team structure and a welcoming experience for the new employee.

Be resourceful and allow talent to thrive. Create a safe but challenging environment where people are free to try new things and even make mistakes. Demand their best work by allowing for errors, but not failure. Errors not corrected become failures. Process Owners should be tasked to learn what they can from their errors and share it with the group. They should be challenged to work hard and always be stretching to improve.

> Demand their best work by allowing for errors, but not failure. Errors not corrected become failures.

Be sure to encourage their hard work and not their intelligence. Intelligence and hard work are not the same thing. Process Owners need to know they can continually work hard on a challenging problem and eventually overcome it. People will give up if they think the problem is too difficult for their intelligence level.

One of the greatest leaders in modern times was Vince Lombardi. As the newly appointed head coach for the Green Bay Packers in the late sixties, Vince led a team of mediocre unproven athletes to win five out of the next seven Super Bowls. There has never been a coach with that kind of record before or since.

Vince never wrote a book about his leadership style but his son did. *How to be Number One*, by Vince Lumbardi Jr., exposes one of Vince's great secrets to motivating his team. Vince understood that people are different and respond to different forms of motivation. Some people like public praise. Others like a private thank you. Some people like pressure, others like hallway conversations. Vince kept a record of what his people responded to and used it to keep them motivated. Learn to read your people and understand how each one of them is motivated.

The Coaching Kata

A Master Chief must be more than a problem solver, they must build Process Owners who can solve problems.

When a Process Owner or a Kaizen Blitz team is stuck on a challenging problem, a Coaching Kata exists for the Master Chief to encourage their progress. As a coach or a mentor, you want to keep them focused. Toyota had a lot of success using Coaching Katas. The Stable Coaching Kata has these questions Master Chief's use while interacting with other in these scenarios:

1. What is the present condition compared to the desired situation?
2. What needs to happen to reach the desired condition?
3. What models exist, or who could we learn from that has already achieved similar results?
4. What is the next step?
5. How soon can learn from the results of the next step?

Coaching is like a marriage. You need to maintain a 5:1 ratio of encouragement verses scrutiny. You must be sincerely interested in your Process Owners ideas, growth, and victories.

Group Problem Solving and Conflict Resolution

Encourage debate, involvement, and critical thinking among the team. Help them understand the best decisions are scrutinized by all sides so that they are made with the least bias. Also, help them understand that measurement removes bias and is an excellent foundation for knowing where we are now and contrasting it with where we want to be.

The Team Development Model

Bruce Tuckman introduced the Team Development model in the mid-sixties which has become a standard in understanding team behavior. When you are starting with a new group of people you don't have a team yet—you have a group of people. You, as the leader, must develop your group of people into an empowered team.

Several stages of evolution transpire as the dynamics within the group evolve from a collection of self-aware people to a single group with a team identity. Bruce Tuckman identified these stages as Forming, Storming, Norming, Performing, and Adjourning.

A depiction of Tuckman's model and an explanation of these stages follow:

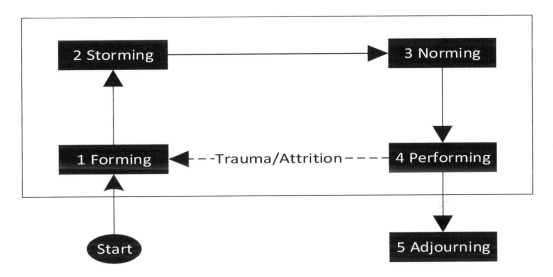

Figure 42 - Team Development Model (Bruce Tuckman)

Forming: At this stage, a group of people meet each other and become acquainted. Introductions are made, and initial first impressions are generated.

Storming: At this stage, the group has been interacting a little and sampling each other's personality characteristics. Loud people are favored in these situations. It's important for you to be heard, and to use relevant industry vocabulary. Experts identify each other within groups by their use of correct vocabulary. Natural leaders, class clowns, quiet thoughtful answer holders are all identified in this stage.

> Experts identify each other within groups by their use of correct vocabulary.

Norming: This stage is where the individuals accept each other and proceed to the next step to start performing as a team, under normal conditions. We'll see in the next step how problems in this step slow down performance in the next step.

Performing: This is where teamwork gets done. Depending on how well the Storming and Norming transitions happened, in this stage the team either performs with full speed, or they limp along, crippled. Sometimes groups don't norm properly and there is unnecessary tension and a lack of trust within the team, or between teams that must work together.

In extreme cases called "Apollo Teams," there are too many Type-A (extremely assertive) people and the team reaches gridlock and cannot seem to get any work done. The key here is vulnerability. A vulnerable team trusts each other and can collaborate better. Vulnerability means each team member feels safe expressing themselves and that they will experience no negative repercussions for it.

Adjourning: This stage is, of course, when the team has completed their work and is disbanding. If it was a long-term engagement you really should have a party or fun memorable event to end the experience properly.

Apollo Teams

In Apollo Team situations, you may need to shuffle some people to other teams or be a 3rd party facilitator when the team needs to make important decisions.

One group I am aware of incorporated a "survivor" theme within their multi-team environment. Team members were empowered to vote each other off of their teams if they contributed to significant problems slowing down the team.

Repairing Team Rivalries

Sometimes in coordinating various teams that must work together, you may encounter two teams who don't work well together. Research was conducted on how to solve this problem in a book called, *The Human Use of Human Resources*. The authors concluded the way to get two conflicted teams to cooperate is to create a series of situations where members on both teams are needed to get the project done. Several scenarios where this approach succeeded are detailed in this book.

The Team Empowerment Model

The Team Empowerment Model is used to develop a foundation for Servant Leadership. Servant Leadership is a leadership philosophy that espouses creating clear objectives, constraints, and intentions, abdicating decision power to the team, putting the needs of others on the team first, and developing people within the team to perform as well as possible. The legitimate leader is then tasked with using his or her authority to remove roadblocks the team may encounter along the way.

As a servant leader, the Master Chief begins by providing the objectives, intentions, constraints, and boundaries for the Process Owners, but then focuses on empowering the Process Owners to be self-directed, champion owners of their processes.

This means the Master Chief pushes the decision making to those closest to the information, letting their team members make the decisions if the decisions don't require extra money or jeopardize anybody's safety or violate any constraints.

Once the team is moving forward, the Master Chief must learn to delegate the workload to the team. This enables the MC to be agile enough to move anywhere within the environment to deal with exceptions as they arise. In other words, the MC delegates the workload and then deals with exceptions as they emerge.

The steps forming the Team Empowerment Model are as follows:

1. Bring Objectives, Constraints, and Dependencies to the Team

The first rule in leading a team is to understand that the objectives for which the team was organized come from outside of the team and should be delivered by the leader to the team. The same is true for any governing constraints. If a team is floundering, not aware of their objectives or constraints, the leader has not yet done his or her job. Some people call this "intent-based leading." or "communicating intent."

2. Challenge the Team to Come Up with the Solution

The newly formed team should then be free to identify how they will achieve their goals. This can be difficult for an inexperienced leader, who hasn't yet learned that ideas multiply when the team is having a healthy dialog. This is a way to push the decision making to those closest to the information.

Depending on the scenario, the MC can play the role of a coach, or a mentor. If the Master Chief happens to be a subject matter expert having experience with an objective and is instructing a new Process Owner, they can take on the role of a coach. A coach helps someone perform better by having inside knowledge and experience with a work item. This obviously adds value to the learning Process Owner.

If the Master Chief has no previous experience with the work item, they then take the place of a mentor. A mentor is different from a coach in that a mentor is an "accountability partner." A mentor does not necessarily have knowledge about the objective but acts as a sounding board, a third-person looking in, and a dependable agent who will challenge the person being mentored to stretch and grow, and to be accountable for commitments made. Much of the work of developing people is mentoring.

Whether coach or a mentor, the Master Chief must play the part of a shaper, challenging the team and individuals to improve. Master Chiefs do this by continually challenging the Process Owners to understand and report on the reality of their environments and then brainstorm ways to improve those environments. The clearer everyone involved understands reality, the better decisions they will make. The better the decisions that get made, the more successful the organization will become.

3. Provide High-Focus Engagement and Support and be Sure the Team is Working Towards the Goal

This is the high-touch stage of leadership. Be present and available to your team. Using the techniques of observation and curiosity, be sure your people have a plan and are working it. Your objective is to build a team of problem solvers, and not be the person solving all the problems.

Don't comment on the validity of their plan. If you have concerns express them in question form. Ask if they are certain it is the best way to proceed? Is it the quickest? Lowest risk? Least expensive? Have they discussed it with others? Is there someone who has already done something similar whom they could ask for advice? What other ideas have they considered? How will they know when they have succeeded?

4. Equip the Team for Success

The fourth responsibility a leader has is to make sure your team has all the equipment and resources needed to succeed. The best way to do this is to ask the team members what else they need to be successful and then act on what they say. Be careful to separate their needs from their wants. Needs should be met as soon as possible and wants can be bargained with. Be sure they are familiar with all the locations for Stable artifacts: The Configuration Management System, Standard Operating Procedures, Kata Card Templates, Asset Recovery Models, Performance Console etc.

If you are unable to provide them with a necessity, explain why and challenge the group to brainstorm a workaround.

5. Delegate, Delegate, Delegate

The fifth responsibility of a leader, once the leader is convinced his or her team knows the objectives and constraints, is equipped for success, and is moving in the right direction, is to delegate the workload to everyone on the team.

There are different schools of thought on this. If you are a Master Chief on a small team then you probably should be busy helping the team do the work. If you have a sizeable team—8-10 or more—you should delegate all the work to your Process Owners and you then stand back and handle exceptions as they occur. This is called "Management by Exception."

If you are busy working on the details every day with a large group, you will miss problem flags and then as these problems become critical issues everyone's work will suffer. Learn to delegate effectively. You will become a better leader and your people will feel more trusted as you delegate and follow-up.

6. Use Positive Psychology

The most important motivating factor an employee can have is to know they are helping another human being. Deming called this "Pride in workmanship." He believed every other form of motivation was a distraction. He was against short-term rewards and motivational slogans. He emphasized the need to establish and elevate the connection between employee and final customer.

Without exception, when managers have complained to me that their employees don't seem to care about their work, they also report that the same employees are disconnected from the end-users their efforts are supposed to help. This is a missing critical success factor in establishing pride in workmanship within a team, and it's the leader's job to ensure this connection is constantly in place. Employees must know who they are helping and have a clear understanding of their customers' needs.

Redirection

In the book, *Whale Done,* Ken Blanchard describes a simple but effective way to train killer whales that works just as well with people. The pattern is called "A-B-C." "A" is the Activator, or goal that a person is challenged to accomplish. "B" is the Behavior that must be observed. "C" is the Conclusion of their actions, being correct or incorrect.

If they are incorrect, the leader performs a technique called "redirection" where the leader steps in and says, "I must not have been clear in explaining my intent. You'll need to do …" and then restates the goal for the employee. This way the employee feels safe to make a mistake, knowing the leader will accept responsibility for the blame.

Ken Schwaber, the co-inventor of Scrum, taught me that when a student in a class doesn't understand something, the instructor should always assume the blame for the lack of understanding. "I'm sorry I confused you," or "I'm sorry I didn't explain that very clearly," and then covering the topic again is an excellent teaching technique.

These techniques breed a culture of safety. People feel like the leader respects them and wants them to succeed.

When complimenting your team, one good technique is to carry around a 3x5 card with all your PROs names on it. Keep track of whom you've complimented recently by putting a mark by each PRO's name when you compliment them. This way you can make sure nobody is neglected.

Although you are trying to keep your compliments stratified, be sure to keep them honest. People can detect fake or shallow compliments.

Wherever you can, compliment your people in front of others whom they respect. This makes a compliment a thousand times more powerful.

Dealing with the Belligerent Employee

Sometimes in rare situations, an employee may demonstrate belligerent behavior that is not acceptable. If they do this publicly, simply announce to them that you will need to see them right after the meeting, or that they will need to come and talk with you privately.

My friend Christian Moore, who has made a career out of speaking to rooms full of educators and prison administrators, has pioneered the channeling of negative energy for constructive use in his book, *The Resilience Breakthrough*. He explains that people's emotions contain both positive and negative energy. Positive energy is easy enough to channel, but it's what people do with their negative energy that separates the geniuses from the felons.

When Steve Jobs was fired from Apple, the company he founded, instead of acting out publicly he took his negative energy and started NeXT Computers, and Pixar. He had so much success, Apple brought him back.

Have a talk with your belligerent employee. Ask them to help you understand what's driving their anti-progressive behavior, and what need they have that is not being met that would make them improve. Be sure they are heard. Listen for whether or not they are being vulnerable to you. If not, they are probably not sharing the root cause of their distress.

Be on their side, but help them re-focus. Challenge them to channel their frustrations into positive results within the environment. If the behavior continues, you may need to work with your leaders and put the employee on a 90-day progressive improvement plan. Council with your Human Resources department for guidance with this. If nothing else works you may have to be firm with them and tell them their behavior indicates they are in the wrong place and if they don't manage their career you will have to do it for them.

Dealing with the Needy Employee

Some employees have learned to take problems that they should be able to solve themselves to their leaders for assistance. If you have an employee like this, simply say to them, "If it were me, I'd research ___ on Google," or "I'd ask _____ how to do that."

Don't do it for them, or you'll disempower them. Remember, one role of a Master Chief is to develop the Process Owners into empowered problem solvers. Jack Welsh, the retired CEO of General Electric, wrote in his book *Winning*, that developing people should be about 40% of a leader's efforts, not 4%.

Dealing with the Irresponsible Employee

Sometimes you may have to deal with employees who blame other people for their inability to get their work done. There is a model called "The Victim Triangle," which is comprised of a victim, a persecutor, and a rescuer.

In the model, the employee presents his or her excuse for failure by blaming someone else or some other factor. This other person, being blamed becomes the persecutor. Your employee is hoping you will become the rescuer by telling them you understand the story and empathize, and that the failure is dismissible this time, but next time the employee should just try harder.

This is dysfunctional. This person learned somewhere in their past that he or she can get rescued from responsibility by having a good enough victim story. A chain is only as strong as its weakest link, and this behavior impacts the whole team negatively.

The way to help mature a person out of this recurring distortion is, instead of empathizing with the situation, challenge them out of this behavior. Point out that the employee's failure hurt the entire team and ask what the employee could have done differently in that same situation to produce a successful result. Challenge the employee to do that in the future if a similar scenario arises. A self-proclaimed victim should be challenged out of any disabling behavior instead of consoled into repeating it.

Dealing with the Challenging Employee

Strange as it seems, you will experience employees who will challenge your authority. This is one of the most difficult behaviors to deal with and is most often experienced when managers are new.

If they ask you a question, it's important that you answer it. You have a lot of options, though. You can solicit the team for an answer to their question. You can answer it yourself. You can also mention it sounds like the employee has a better idea and you'd like to hear it.

Be sure to express yourself from the perspective of what's best for the customer, the sponsor, the work product, or the team. These perspectives are always safe territory.

If your judgment is challenged make a firm effort to explain clearly why you made the decision.

Dealing with Conflict Within the Team

Whenever there is conflict within the team, always focus on the problems, and not the people. You may have to remind your teams to keep that focus.

In addition, teach them to state how their opinions would help the customers, or the project, or the company, or the team. If they can learn to state their opinions from those perspectives, everyone will be on safe ground.

7. Lead Them to Reality

Measurements reflect reality, and part of a Master Chief's role is to challenge the team to create effective measurements to indicate their progress. Measuring should not be used to reward or punish, but rather to show reality.

After measurements are available, the team can brainstorm better ways to improve and even better types of measures. Measuring is a window into reality and the secret to accountability and improvement. What gets measured gets done, and can be improved.

8. Remind the Teams to Systematize so that they can Learn from Their Past

Your teams will be busy brainstorming, interfacing with customers, completing Kata Cards and SLAM Charts, performing process work, and getting ready for the next Master Cycle. They will need you to continually remind them of continuous improvement activities until the culture shifts enough that it becomes second nature for them.

9. Laugh and Have Fun

Before people studied workforce behavior, laughter was thought of as wasting time. Management projected a facade of somberness and stern personalities to keep workers focused.

Companies are discovering laughter has amazing benefits for the workplace. Among other things it reduces stress, reduces resistance to change, increases morale, disarms anger, improves attitudes, breaks the ice, increases trust, and improves team cohesion.

Have fun together and create positive memories as a team. It will make the workplace a welcoming environment.

Three Models of Excellent Teams

Three models of excellent teams every Master Chief needs to be familiar with are published in three different books.

In Jim Collins book *Good to Great*, he talks about the Hedgehog Concept. This concept is where a team is characterized by having three consistent qualities in alignment with their work. These consistencies were found in almost every world-class company Collins' team of researchers studied. The questions that identified these qualities were:

1. What are you passionate about?
2. What drives your economic engine?
3. What can you be best at in the world?

In most cases, every world-class company felt passionate about its work and felt it could be world-class at it with the right management support.

In Marcus Buckingham's book *First Break All the Rules: What the World's Greatest Managers Do Differently*, Buckingham's Gallop Poll team discovered that top performing companies had employees that for the most part could answer yes to the following twelve critical questions. Ask your team these questions and ask for feedback on how to bridge any gap. The questions are quoted from the book:

"

1. Do I know what is expected of me at work?
2. Do I have the materials and equipment I need to do my work right?
3. Do I have the opportunity to do what I do best every day?
4. In the last 7 days, have I received recognition or praise for doing good work?
5. Does my supervisor or someone at work seem to care about me as a person?
6. Is there someone at work who encourages my development?
7. At work, do my opinions seem to count?
8. Does the mission or purpose of my company make me feel my job is important?
9. Are my co-workers committed to doing quality work?
10. Do I have a best friend at work?

11. In the last 6 months, has someone talked to me about my progress?
12. This last year, have I had the opportunity at work to learn and grow?

"

Finally, Google spent two years studying what makes a great team and concluded the following five factors are critical:

1. Psychological Safety – The ability to be honest, and even vulnerable with other team members without repercussions.
2. Dependability – The team could depend on each other and had confidence that work items would get completed by their teammates when expected.
3. Structure and Clarity – The teams had clear goals and constraints.
4. Meaning – The work had personal significance for each member.
2. Impact – The group believed their work was purposeful and that it positively impacted the greater good.

Are you beginning to see the pattern? Employees in all these groups work in an environment of respect, trust, vulnerability, purpose, empowerment, structure, respect, visibility, and success.

The Future Domain and the Master Chief

A Master Chief starts the Quality Planning process when the Stable Framework™ is first implemented by establishing the structural components within the environment. Every six months or so the Quality Planning process should be revisited.

During Quality Planning, The Master Chief is responsible for establishing the Configuration Management System and all the structures inside of it. Next, the Master Chief creates the System Schedule and begins populating everything that must go on it.

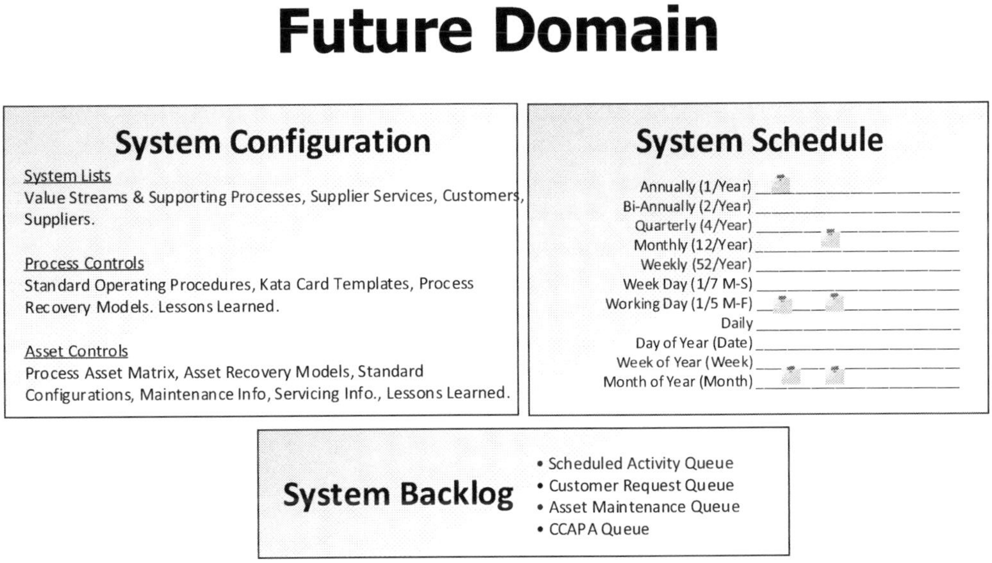

Future Domain

System Configuration

System Lists
Value Streams & Supporting Processes, Supplier Services, Customers, Suppliers.

Process Controls
Standard Operating Procedures, Kata Card Templates, Process Recovery Models. Lessons Learned.

Asset Controls
Process Asset Matrix, Asset Recovery Models, Standard Configurations, Maintenance Info, Servicing Info., Lessons Learned.

System Schedule

Annually (1/Year) _____
Bi-Annually (2/Year) _____
Quarterly (4/Year) _____
Monthly (12/Year) _____
Weekly (52/Year) _____
Week Day (1/7 M-S) _____
Working Day (1/5 M-F) _____
Daily _____
Day of Year (Date) _____
Week of Year (Week) _____
Month of Year (Month) _____

System Backlog

- Scheduled Activity Queue
- Customer Request Queue
- Asset Maintenance Queue
- CCAPA Queue

Figure 43 - The Future Domain

Finally, the Master Chief creates the Customer Request Queue, Scheduled Activity Queue, Equipment Maintenance Queue, and the CCAPA Queue.

In addition, the Master Chief must establish the Performance Console and the historical logs for retaining past information over time.

The Master Chief assumes the role of the champion of process innovation in the Future Domain. Master Chiefs are always challenging the Process Owners to improve the results and efficiencies of their processes. They own the System Schedule but work with the PROs to put their scheduled events on it. This is done in the Master Cycle Planning Meetings.

Once the team is up and running, as the Process Owners receive customer requests from outside, they place them in the Customer Request Queue, or in the Objectives column on the Kanban board if they are urgent.

As activities listed in the System Schedule become imminent, the Master Chief puts them into the Scheduled Activity Queue. This should happen at the beginning of each Master Cycle Planning Meeting.

All items added to the queues should be categorized and then timestamped. They should be timestamped repeatedly as they move through each column on the Kanban. This provides empirical information for estimating lead times for future similar activity requests.

Nonconforming product or service problems that are experienced by customers go into the CCAPA Queue for review and action. These problems are brought to the team usually from the Service Desk. The Service Desk should pool common problems and bring them to Engineering or Operations.

The CCAPA Committee is a combination of representatives from the Service Desk, Engineering, and Operations who receive non-conforming problems experienced by customers and then prioritize them for remediation based on impact. Some groups hold CCAPA Committee meetings daily, and some hold them weekly or once per cycle. You might consider holding daily meetings after a major release and then weekly or once per cycle until the next major release.

The Present Domain and the Master Chief

The Master Chief sets up the Kanban Board and the recurring schedule for the CCAPA and Change Control committees during Quality Planning.

Present Domain

Team
- Daily Kaizen Standup
- S5 Housekeeping
- Adhoc Kaizen Teams
- Process Kata & Kata Cards
- Training/ReTraining

Committees
- Change Control Committee
- CCAPA Committee

Environment
- Alert System
- Production Change Log
- Suggestion Box

Kanban Board

Objectives	Selected Tasks To Do	In Process (3)	Done

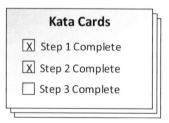

Kata Cards

[X] Step 1 Complete

[X] Step 2 Complete

[ ] Step 3 Complete

Figure 44 - The Present Domain

In addition, Master Chiefs coordinate the Master Cycle Planning Meeting, the Daily Kaizen Stand-up meeting, and the Cycle Review and Cycle Retrospective meetings.

Master Cycle Planning Meeting

At the beginning of each Master Cycle, the Master Chief calls the Cycle Planning Meeting, which he or she attends with the Process Owners. In this meeting, they examine the System Schedule and the Work Queues in the System Backlog and load any urgent items from the work queues into the Objectives column of the Kanban board. The MC moves the most critical items to the top.

Depending on the environment, the MC can meet with a representative group from the Service Desk to discuss any new nonconforming issues that were reported from the customer base since the last meeting. Some groups do this every morning before the daily Kaizen Stand-up Meeting, and some groups just do this as an agenda item during the Master Cycle Review Meeting.

The MC should also be getting updates from Engineering regularly so that the MC is aware of any new additions or changes to products or services. The Master Chief could attend the Engineering Status Meetings or invite an Engineering representative to their Cycle Review Meetings.

If there is a project team involved, they can come prepared with a Gantt Chart, if needed, and a Product Backlog with activities to load into the "Selected Tasks To Do" column.

Daily Kaizen Stand-up Meeting

The Master Chief calls the daily Kaizen Stand-up Meeting. During this meeting, the MC spends five minutes updating the group about changes in the marketplace and organizational news, if they have changes to report. The MC is a communication conduit from Management to the team.

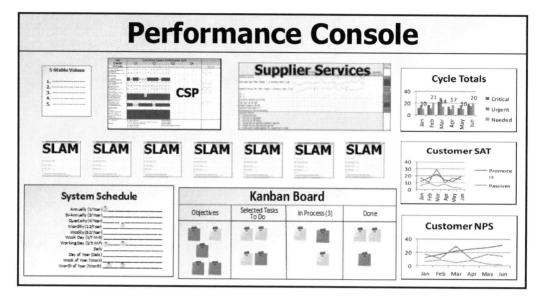

Figure 45 - The Performance Console

The team meets around the Performance Console for their Stand-up Meeting. The performance console is the centerpiece for collaboration and communication.

The Performance Console must be highly visible, and information on it should be updated regularly and be easy to read. Feel free to add additional components to it such as a team morale report, employee of the week, personas, or daily comics.

The MC reads any new suggestions from the suggestion box. Innovation and improvement should be encouraged and should be a daily conversation.

The MC should take suggestions seriously. Taking quick action and demonstrating follow-through on suggestions helps grow trust, and the absence of this action erodes it. If a suggestion cannot be implemented the MC needs to explain why to the team.

The MC is free between or during the daily Kaizen Stand-up Meeting to reprioritize items in the "Objectives" column of the Kanban board. Team members are also expected to add new items to the Customer Request Queue, or the "Objectives," or "To-Do" column as needed. After the MC has prioritized the items to be completed, the team then self-selects the new, more urgent items. In some cases, it may be necessary to place items from the "Objectives" column back into the work queues.

Process Owners on the team update the status of their work items on the Kanban Board and choose from the "Objectives" column a new objective and activities. If the objective has not already been broken down into tasks, the Process Owner breaks it down into a list of two day or smaller tasks and populates the "Selected Tasks To Do" column with the decomposed list of tasks.

Change control can happen in the daily Kaizen Stand-up Meeting, as everyone on the team should be present for signatures, and to express concerns. If a long conversation is needed to discuss a concern, the people involved can postpone the conversation until after the meeting is finished. In some situations, you may want to just have change control during the Cycle Review Meeting.

The Master Chief also makes sure that every production change made is recorded in the Production Change Log. This log becomes priceless when searching for root causes of a production problem.

The Past Domain and the Master Chief

The Master Chief is responsible for ensuring the PROs update their

Past Domain

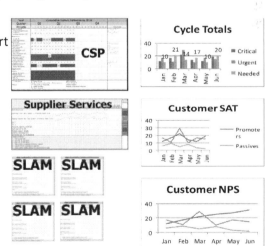

Master Cycle Totals
- Cumulative System Perf. (CSP) Chart
- Supplier Services SLAM Chart
- Cycle Totals
- Customer Satisfaction
- Customer Net Promoter Score
- SLAM Chart A, B, etc.

Historical Logs
- Production Change Log
- Critical Decision Log
- Kata Card Historical Repository
- CSP Repository
- Training Log

Figure 46 - The Past Domain

SLAM Charts at the Daily Kaizen Stand-up Meeting, and the Cumulative System Performance Chart at the end of each Master Cycle.

They are also responsible for populating the Supplier Services SLAM Chart daily, and updating the Cumulative System Performance Chart with that data at the end of each Master Cycle.

This helps the tracking of long-term process capability trends within the System. This gives the department sponsors a timeline view of how the value streams in the system are performing, which indicates the health of the overall system.

Process Owners should always be challenged to improve their processes. Two starting points for improvement are increasing the target customers satisfaction levels, and then improving the ability of the company to deliver its services more efficiently by making fewer mistakes and consuming less time, money, and other resources.

Chapter 14 - Committees, Groups, and Kaizen Events

Of course, organizations are made up of many functional units. Whether you are in Operations, Implementation, DevOps, or Development you're sure to be interacting with the other functional units within your organization. The explanations that follow are models for how these interactions can happen.

Change Control Committee

As production changes are proposed, they may impact other processes going on around them. If these requests are standard recurring production changes, typically the change has been systematized and requires only a notification, not an approval. However, new changes to production or even some systematized changes must be understood by all concerned and approved to go forward.

In Stable, your whole team is the change control committee and is made aware of any pending changes during the morning Kaizen Stand-up Meeting. They are free to put a hold on a pending change if needed until they can adjust their own environment and approve the change.

If your team interacts with other IT groups, you may have a weekly multi-department Change Control Meeting. You simply queue up your changes every day for that weekly meeting.

If your whole department is practicing Stable and you are having a set of escalated morning Kaizen Stand-ups (Kaizen of Kaizens) then you can push changes daily by bringing the approved changes in the early Kaizen Stand-up Meetings to the Master Kaizen Meeting.

CCAPA Committee

CCAPA is an acronym that means Correction, Corrective Action, and Preventative Action. These three actions are what happens when a Nonconforming Incident is reported by a customer.

A Nonconforming Incident is any incident where the customer experiences something that wasn't as it should be. You can think of your

Service Desk receiving calls and logging a problem as receiving a report of a nonconforming incident.

These incidents, if recurring or significant, should be passed to the CCAPA committee. The CCAPA Committee is then tasked with the following:

- Correction - Making sure the situation is corrected for the specific customer or customers that reported it, if the Service Desk has not already done so.

- Corrective Action - Identifying what should be adjusted in the current Process Controls (SOP & Kata Cards) to prevent this type of situation from happening in the future, and then making sure those adjustments are made.

- Preventative Action - Identifying any other processes in the environment that could benefit from a similar adjustment, and them being sure those adjustments are made.

If you are operating in a controlled environment, such as an FDA regulated industry, you'll need to keep a record of all these CCAPA issues and resolutions.

Daily Kaizen Stand-up Meeting

The Daily Kaizen is a daily morning Stand-up Meeting called by the Master Chief. The meeting is attended by the Master Chief and all Process Owners, or those on the shift.

The meeting should last no longer than 25 minutes. The meeting consists of the following:

- The Master Chief optionally provides a brief 5 minute or less market status report which is a summary of any outstanding news from Senior Management or the Service Desk or any other part of the company. The Master Chief may want to meet regularly with the Service Desk (CCAPA Committee Meeting) for this purpose before the daily Kaizen each morning, or once a week.

- The Master Chief notifies the team of any new items or items that have received a higher priority.
- The Process Owners clarify any new work items they have added to their "Selected Tasks To Do" column.
- The team updates their SLAM Charts with yesterday's information and reports on the following four questions:

    1. What did I accomplish yesterday (including any change control items)?
    2. What do I intend to accomplish today?
    3. What roadblocks are in my way?
    4. What did I improve in my environment since yesterday?

Kaizen Teams

Kaizen Teams are groups of team members organized to solve problems. There are two reasons to form a Kaizen Team. A team many be formed to work on a Kaizen Project during a Master Cycle, and a team may be formed ad-hoc at any time to assist another Process Owner or a Master Chief with an urgent problem. We call this situation a Kaizen Blitz.

Toyota discovered that volunteer team members perform much better than assigned members. Toyota also learned that encouraging hard work instead of intelligence for problem solving brought better results. If intelligence was encouraged, people would sometimes think a problem was too difficult intellectually to solve and they would give up. However, if they were encouraged to work hard on a problem they would usually eventually succeed. Use volunteers whenever possible and be sure to encourage hard work and persistence over intelligence.

Kaizen Project

During each Master Cycle the team should be encouraged to pick an item in their environment that needs to be improved, and work individually on their own items, or together on the same item. Each challenge we call a Kaizen Project. Examples of Kaizen Projects include creating an automation script, a customer survey, a useful web post, or maybe organizing information better.

Kaizen Blitz, or Ad-hoc Kaizen Team

A Kaizen Blitz is an ad-hoc team formed when a Master Chief or a Process Owner requests volunteer assistance with a problem. This can happen any time during a Master Cycle. Team members should be encouraged to ask each other for assistance.

Master Kaizen Meeting, or Kaizen of Kaizens

When implementing Stable in a large environment, you will have several Stable teams, all requiring their own Kaizen Stand-up Meeting. This works fine as long as no one person is on multiple teams. If this is the case, however, you can stagger the meetings from, for example, 9:00 to 9:30 to 10:00, etc.

Alternatively, you can have a set of Kaizen Meetings at 9:00, and then a Master Kaizen Meeting at 9:30 where a representative from each team reports at the later meeting. This Master Kaizen Meeting is called a Kaizen of Kaizens.

Gemba Walk

"Gemba" in Japanese means the location where the work gets done. That means a team has their own Gemba, and the customers have their own Gemba.

Sometimes the best way to understand a problem, or get clarity on a requirement, is to visit customers in their own Gemba. We call this a Gemba Walk. It ensures a deeper understanding of the work or issue resulting in a better decision than can be had from just staying inside a conference room.

Chapter 15 - The Stable, or "Hybrid" Project

Stable projects are typically Operations or Implementations efforts and tend to have many repeatable characteristics. In addition, these types of projects often have known dependencies that must be identified during planning. For this reason, Stable projects combine the best of Plan-based and Agile project management approaches into what is commonly called a Hybrid Project approach. A good understanding of the strengths and weaknesses of these two basic approaches is helpful for understanding how to plan and execute a Stable Project.

Challenges that require project management emerge in two categories of disorder: Disorganized Disorder, and Chaotic Disorder.

Disorganized Disorder is a situation where a team is tasked with building something similar to what has been built somewhere in the past. Precedents exist, the team just needs to build something similar. Projects like real-estate construction, road construction, copying competitive products, or even reproducing a new software application on a different technology platform are all examples of disorganized challenges. Plan-based project management works best for these types of challenges because the team is dealing with relatively well-known steps and outcomes, and typically has a lot of past examples to refer to as needed.

Chaotic Disorder is a situation where the project team must build something that has never been built before. The solution will be original because there is nothing like it previously existing. The team doesn't have all of the answers yet and must learn as they go. Agile project management approaches work best for these situations.

In summary, Disorganized Disorder is best addressed with a Plan-based approach, while Chaotic Disorder is best addressed with an Agile approach. Let's take a look at the differences between Plan-based, and Agile approaches.

Plan-Based Project Management

It is common in the industry to refer to plan-based projects as "waterfall" projects.

Plan-based project management as we know it today started at the beginning of World War I when the US Army hired management consultant Henry Laurence Gantt to help with supply line logistics.

His answer was a tool empowering a project team to identify deliverables, list all known activities required to complete each deliverable, estimate durations required to complete each activity, and then identify dependencies between activities so that the plan could be carried out efficiently.

Years later, in 1969, the Project Management Institute® was created and grew to become the primary authority on plan-based project management. Today they offer the PMP© and CAPM© project certifications, which are excellent sources for learning about executing projects where the result is an output that can be easily modeled after something that already exists. This allows a skilled team to estimate the overall schedule and cost of a project by understanding work items based on past similar projects.

Plan-based projects require decomposing the entire known workload into a set of distinguishable deliverables called a Work Breakdown Structure, and then creating a Project Schedule and Gantt Chart listing the activities required to complete each deliverable. All known activity dependencies can be identified to ensure the project can be executed in the correct order. This approach to project management is traditional in many companies and contains the reporting components that senior management is often use to seeing.

The drawback to this approach is that if your team is tasked with producing a solution based on a chaotic challenge, the plan will change so often that it will age into irrelevance quickly. After many failed software projects, the industry responded with Agile.

Agile Project Management

To facilitate hyper-changing environments and unknown situations, Agile Project Management was envisioned in 2001. Agile emphasizes short delivery cycles of high-focused teamwork producing deliverables of maximum value to the customer for a minimum effort on the team's part. Regular customer demos at the end of these cycles—called Sprints—

ensures the target deliverables stay relevant as they are being developed.

Instead of creating a thorough plan up front, objectives are kept high-level on a list called a Product Backlog, and team members select the highest priority objectives to complete during each Sprint. At the end of the Sprint, the team re-groups, re-evaluates the priorities, and selects the next group of highest priority objectives to work on during the next Sprint. These techniques allow the team the flexibility to react almost real-time to changes within the environment, to learn as they go, and to produce direct value as quickly as possible for the customer.

We found in Chapter 2 that after Agile was introduced into the software industry in 2001, the improvements in the years following that indicate how project success rates have gone up about nine percent, and the rate of failed projects have gone down about nine percent, without really impacting the late project statistic.

While failed projects tend to happen because the result was not a good fit for the customer base, late projects tend to happen because of the Hidden Factory. So, as an industry Agile has transformed our failed projects into greater successes, but hasn't really addressed the Hidden Factory in software. In addition, the drawback of Agile project management is that it can be challenging to identify all of the components, dependences, and the total schedule or cost requirements for a project up front. This is where Stable helps.

The Stable Project Management Approach

Stable combines the best of Plan-based and Agile project management approaches to produce an outward facing project structure resembling a traditional project, but an inward facing Agile engine which is most efficient for the team.

Stable can work in harmony with your current role structure. Existing Project Managers or Agile Product Owners can keep their titles and simply wear the hat of a Process Owner as they work their Stable projects.

In Stable, we use a Project Schedule & Gantt Chart tool as our Product Backlog. This allows us to manage dependencies and to use the additional columns to add helpful traceability data to each item such as what project goal the item supports, which component it will be found in, who is the subject matter expert on the item, and in which test case it was confirmed as working properly.

Items in the Product Backlog represent deliverable components and all of the activities required to complete each component. It's best to show activities rolling up into components. Additional scope can be added to the Gantt Chart as needed and then sized and prioritized accordingly.

The Project Manager continually moves the highest prioritized items from the Backlog onto the top of the "Objectives" column of the Kanban Board where the team self-selects who will complete the next most urgent objective. As the objectives are completed, progress is updated on the Gantt Chart. This way dependencies are respected, sponsors can see project progress, and the team is insulated from dependent concerns. A Burn-up Chart with a total-scope series line provides further transparency on team velocity, total scope identified, and work completion trends.

Following is a detailed look at the Stable project approach.

Starting a Stable project

Stable projects start with a process called Chartering, where the Project Manager captures important summary information about the project in a one-page Project Charter. Typical pieces of information in a Project Charter include the following:

- The project sponsor
- The project manger
- The project problem needing to be solved
- Primary goals and success criteria
- Any obvious objectives that comprise the primary goals
- Primary stakeholders and roles in the project community
- Any noteworthy risks, constraints, dependencies, or exclusions
- Approval via sponsor signature

If a more formal project plan is needed, such as when a group is working in a regulated industry, a Standard Operating Procedure can be created explaining how project work is performed, and this SOP can be referred to in the Project Charter. This chapter may serve as the foundation for an SOP.

Identifying the Project Scope

If you have no previous project template to model your project after, your team may need to first create a Work Breakdown Structure, and then transfer the resulting work packages as entries into your Product Backlog.

An excellent way to generate a Work Breakdown Structure is to use an Affinity Diagraming technique. A Define Scope meeting is called, comprising the project sponsor, the team, and one or more representatives from the user-base. Participants take 3x5 cards and write one deliverable or feature per card. After everyone has finished writing, the cards are collected. A facilitator is chosen to organize the cards into similar groups. Duplicates are eliminated and the result is a Work Breakdown Structure.

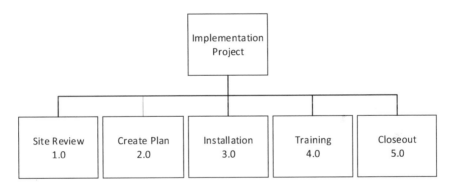

Figure 47 - Work Breakdown Structure (WBS)

Next, deliverables are transferred from the Work Breakdown Structure into the Product Backlog (Schedule and Gantt Chart). Working together as a team, all dependencies between deliverables are identified, and then all the activities required to produce each deliverable are decomposed into two day or smaller sizes. Each day represents an effort point, or stone. Add up all the stones and the team has the total amount of stones of effort estimated for each objective. Collectively, they comprise the total stones of effort required for the entire project.

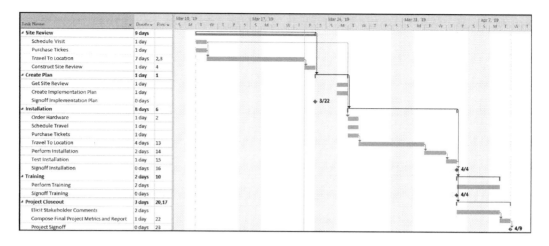

Figure 48 - Project Schedule & Gantt Chart

If the project scope is large the Project Manager and team may need to spend time interacting with individual project stakeholders collecting requirement information about what they need. Storyboarding or Use Case generation activities are excellent techniques for identifying user roles, and their needs. In addition, the team should perform enough of the design work to determine the architectural components required to fulfill all of the identified needs.

The information collected would then be taken to the sponsor so they can decide how much of the requested scope they are willing to fund. End users will give you a long wish list and the sponsor may not want to pay for everything. Be sure to separate the needs from the wants when identifying requirements.

To save time, requirements can be written in a testable manner, so that they double as test plans, eliminating much of the work needed to create tests for the project. For ambiguous requirements, a standardized sentence has emerged in the industry which is useful for capturing periphery information about a requirement. This can be used when needed, for clarification. It is:

"As a [user type] I need to [do something] for the purpose of [stated objective]."

We learned from Agile to keep requirements to a minimum in the beginning, and then elaborate them as we go. It's important to identify the user roles, their needs, and any needed architecture components, but the specific details of the requirements of each item can be identified later during development unless it has unusual characteristics. Anything ambiguous or complex should be scoped and sized up front. The remaining items ought to be reasonably estimated by the team based on past experience.

Later, you can track the cycle velocity of each team by counting how many stones of effort get completed per Master Cycle, and you can estimate the total length of time required to complete the project, based on past cycle velocity.

Executing the Project Work

At the beginning of each Master Cycle, the Project Manager gets with the team and puts the highest priority objectives from the Product Backlog in to the top of the "Objectives" column of the Kanban Board. The "Objectives" column of the Kanban Board can be thought of as the top of the Product Backlog.

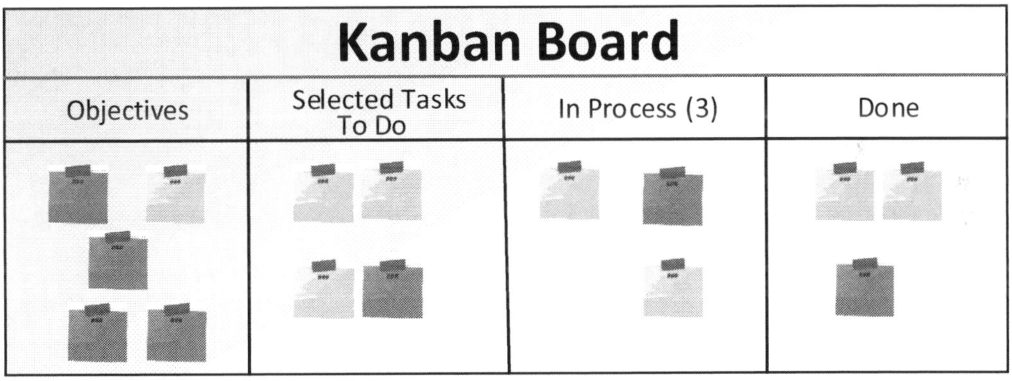

Figure 49 - Project Kanban Board

The team then starts off individually selecting which item from the top of the "Objectives" column they want to accomplish first. Unlike Agile, no commitments are made for completion, but instead, the team understands that when they have completed these items, they are to select the next items from the top of the "Objectives" column.

As the "Objective" column decreases in size, the Project Manager can add to it at any time, keeping it queued. New items can be discussed during the morning Kaizen Standup meeting, as needed.

As daily production needs interrupt scheduled work, the team can add the new items to the "Selected Tasks To Do" column, and then work to complete them. When finished, they can return to the next planned work item in the "Objectives" column.

Towards the end of the Master Cycle, instead of selecting new items that may not be completed within the cycle, team members may assist other members to get "In Process" items completed by the end of the cycle.

At the Master Cycle Review meeting held when the Master Cycle is over, the team demonstrates what they have produced to the Sponsor and any additional stakeholders invited to the meeting.

Reporting on Project Progress

Many companies have developed a standard report for showing the health of all ongoing projects in a Red-Amber-Green stoplight format. In Stable we call this the Master Project Schedule. An example of this chart is shown in the "Stable Portfolio Management" chapter in this book.

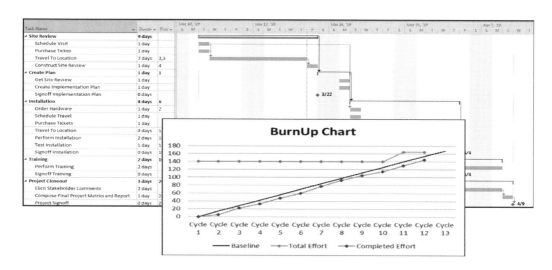

Figure 50 - Reporting Project Progress with a Gantt and a Burnup Chart

For detailed information about a project, the team can report using a Gantt Chart and a Burnup Chart. The Gantt Chart shows a snapshot in time. The Burnup Chart represents the history and work performance trends. The cycle velocity, or amount of stones completed each cycle, will form a trend where the project completion date can be estimated using projection.

If needed, it may be helpful to create a pictorial representation of the deliverables as a Work Breakdown Structure and use it to report on progress, division of labor, or cost breakdowns.

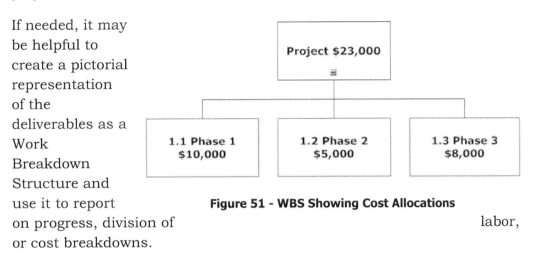

Figure 51 - WBS Showing Cost Allocations

Risk Management

In an Agile-like manner, Stable is able to handle risks naturally, as they occur. In addition, Kata Cards decrease the number of problematic events over time as defensive institutional knowledge grows within the project environment. As we work on projects, we have the added benefit of being able to anticipate possible future risks and plan accordingly.

Any risks identified during the Chartering session, or by examining past similar project histories, or throughout the duration of the project, should be listed on a Risk Register and posted somewhere publicly, preferably on the performance console. Risks should be revisited and discussed at each Master Cycle Planning meeting.

The Risk Register should be one of the final documents stored with the final project report. This way, the list of risks can inspire future Project Managers executing projects in a Stable environment.

Improving Team Performance with Retrospectives and Kaizen Projects

After the Master Cycle Review meeting, the team excuses the non-team stakeholders, thanking them for their time. The team now begins their Master Cycle Retrospective meeting. The Retrospective Meeting is called by the team facilitator.

In this meeting, the team askes themselves these three questions:

1. What went well this past cycle?
2. What could have gone better this past cycle?
3. What still feeds the hidden factory?

A discussion is had, and takeaways are used to improve the environment. As Process Owners, they can identify a small Kaizen project they can work on together, or individually, to improve their efficiency.

Closing a Project

As a project is wrapped up, it is important to preserve important information collected during the experience for the benefit of future projects.

A Final Project Report containing the following information should be filed away with the project in a location easy to access for future Project Managers. The report should contain:

- Comments from primary stakeholders – What went well? What could have been better?
- Lessons learned.
- Risks identified and mitigation plans.
- Estimates vs. actuals vs. actuals adjusted for overtime.
- Amount of additional scope that was added during the project. Different sponsors have different coefficiencies in this area.

Continuous Improvement with The Stable Framework™

Finally, it's important to use Stable Framework components to improve your project efforts over time.

Create a set of project Kata Cards. Example Kata Checkpoints include the "Requirement Review," "Design Review", "Code Review" for Quality Assurance, and then "Smoke Tests," "Component Tests," "Integration Tests," "System Tests," and "User Acceptance Tests" for Quality Control.

An important tool to create and improve as each project completes is a department "Estimation" Standard Operating Procedure and Kata Card. Start simple, but use it each time a new project begins, and improve it every time a project is completed. Examine the estimates verses actual numbers. Be sure to correct for overtime. Have a Project Retrospective at the conclusion of each project where your team brainstorms how the Estimation Procedure and Kata Card can be improved.

Chapter 16 - Using Stable in Operations

Operations is the daily support of existing products or services. In IT, Operations is considered anything from network support to running batch jobs at night, to monitoring large websites, to the administration of layers of middleware and databases. Backups and antivirus and disaster recovery fall under this area, too.

Operations tasks are thought of as recurring on a scheduled basis. A typical Operations model is the Schedule-Do-Check-Adjust Cycle, shown in the figure below. In this model activities identified to be completed are scheduled for the end-of-business-day, or end-of-month, or whenever needed, and then started. This is the first step, "Schedule."

It's common for an Operations group to have a combination of automated and manual tasks. The automated tasks should have a historical log indicating they completed successfully, and an alert mechanism when they don't. The operators will know if the manual activities completed successfully or not because they are performing them in real-time. This is the "Do" step.

If no alert mechanism is in place, the operator must check to be sure the actions completed successfully, which is the "Check" step. If the activity abended, it is examined and re-run, usually with some minor adjustment.

If everything worked as planned, the cycle repeats itself. If something needs an adjustment, the process is adjusted in the "Adjust" step.

This process repeats itself and evolves as the organization's IT needs grow. Two areas that often get neglected in Operations are scheduling regular Disaster Recovery tests and scheduling regular maintenance on Operational assets.

Don't be that shop where you have a Disaster Recovery plan, in theory, but it never gets tested until a real emergency. Test it regularly, perhaps every six months over a long weekend.

Also, learn to think of your assets as a fleet, and schedule routine maintenance for them. Every so often, defrag all the server drives, check average CPU saturation over time, check swap file size to RAM size ratios, check all firewalls, and get rid of any unnecessary apps or fancy background desktop graphics that make servers run slower. Over time, with a lot of operators, people get in and mess with stuff. Be proactive.

Operations Model

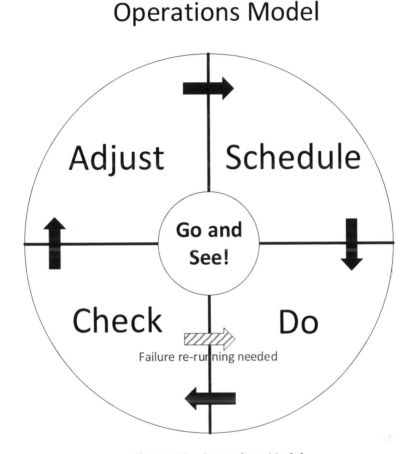

Figure 52 - Operations Model

Operations Using the Stable Framework™

For an Operations group, a quality program is critical. Repeating processes must work the first time, every time. A Systematized Framework like the Stable Framework™ fulfills this need.

We start by identifying all value propositions, and then all repeatable processes within their value streams, scheduling them on the System Schedule, and adding their information to the Master Lists and Process Controls in the Configuration Management System.

Next, we identify our appropriate Master Cycle length which we will use for short-term planning and reporting. A Master Cycle should span anywhere from one to four weeks and should match an Agile Sprint, if other teams in the organization are working in an Agile environment. This ensures reporting and planning happen together, and inter-department change control can be managed easier.

Then, during the Master Cycle Planning Meeting at the start of our next Master Cycle, we identify everything that needs to happen during that next cycle, as listed in the System Schedule.

The Master Chief puts the most urgent items into the "Objectives" column of the Kanban board, ordering them by most to least urgent, and the team self-selects which items they will work on next. The team then decomposes the items into two day or smaller tasks, moving them into the "Selected Tasks To Do" column of the Kanban board.

The team will naturally have their own requests to fulfill during the work day. They should indicate these in the "Selected Tasks To Do" column as they begin working on them. As they finish their own personal backlogs in the "Selected Tasks To Do" column, they can pick a new item from the "Objectives" column.

Repeatable processes that are performed manually require Kata Cards. Kata Cards can be thought of as super-checklists. They contain any steps needed to prevent problems from occurring within the process and are based on knowledge gained from past problems that have occurred. They may also collect information, if needed. When the process is complete, these Kata Cards are passed down to the next Process Owner receiving the work product, as a certificate the work was completed correctly. The final Process Owner gives the stack of collected Kata Cards to the Master Chief for storage in a historical log.

As the team is busy performing their tasks during the cycle, team members meet every morning for a daily Kaizen Stand-up Meeting to review the work at hand and anything new they or the Master Chief may need to add to the list of prioritized objectives.

Optionally, the Master Chief spends five minutes giving the team a "State of the Market" report, and team members spend 20 minutes communicating with each other what they completed yesterday (including change control items), what they intend to accomplish today, if they have any bottlenecks, and what improvements they have made in their workplace. The change control items can be signed at that time, if necessary.

Actual process work being performed during Operations comes from the System Backlog, which is a combination of the Scheduled Activity Queue, the Customer Requests Queue, the Asset Maintenance Queue, and the nonconforming production issues that become Corrections, Corrective Actions, and Preventative Action (CCAPA) items. All of these can be tracked using the Kanban Board, and may require the use of Kata Cards for error-proofing and accountability.

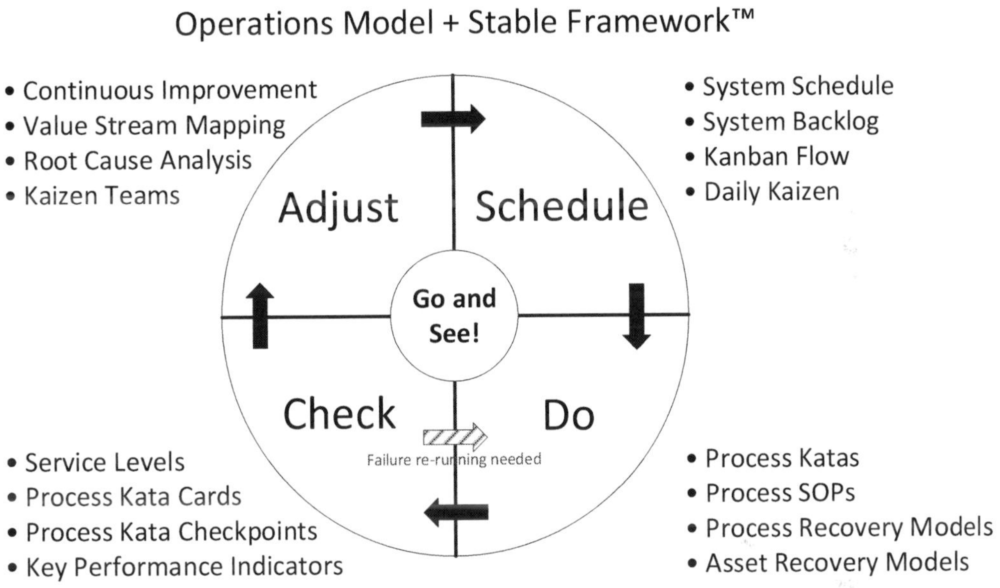

Operations Model + Stable Framework™

- Continuous Improvement
- Value Stream Mapping
- Root Cause Analysis
- Kaizen Teams

- System Schedule
- System Backlog
- Kanban Flow
- Daily Kaizen

Adjust | **Schedule**

Go and See!

Check | **Do**

Failure re-running needed

- Service Levels
- Process Kata Cards
- Process Kata Checkpoints
- Key Performance Indicators

- Process Katas
- Process SOPs
- Process Recovery Models
- Asset Recovery Models

Figure 53 - Operations Model + Stable Framework™

Operations should have agreed-upon levels of performance, known as Service Level Agreements (SLA's). If no contract stipulates an SLA, then a Service Level Goal (SLG) can be created with the business sponsors. These are pre-established targets for performance set by the Master Chief or Process Owner while consulting with the customers of the service.

We demonstrate our ability to meet these service levels by showing a Service Level Attainment Monitor (SLAM) Chart in a visible location called a Performance Console.

These SLAM Charts can be updated daily and reported on at regular intervals called Master Cycle Review Meetings.

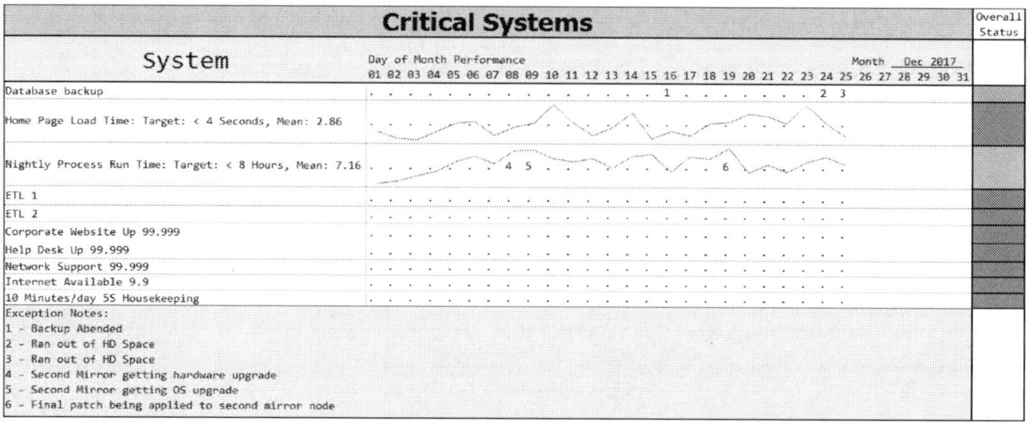

Critical Systems			Overall Status
System	Day of Month Performance Month Dec 2017		
	01 02 03 04 05 06 07 08 09 10 11 12 13 14 15 16 17 18 19 20 21 22 23 24 25 26 27 28 29 30 31		
Database backup	. . . . 1 . . . . 2 3		
Home Page Load Time: Target: < 4 Seconds, Mean: 2.86			
Nightly Process Run Time: Target: < 8 Hours, Mean: 7.16	. . . . 4 5 . . . . 6		
ETL 1	. . . .		
ETL 2	. . . .		
Corporate Website Up 99.999	. . . .		
Help Desk Up 99.999	. . . .		
Network Support 99.999	. . . .		
Internet Available 9.9	. . . .		
10 Minutes/day 5S Housekeeping	. . . .		
Exception Notes: 1 - Backup Abended 2 - Ran out of HD Space 3 - Ran out of HD Space 4 - Second Mirror getting hardware upgrade 5 - Second Mirror getting OS upgrade 6 - Final patch being applied to second mirror node			

Figure 54 - SLAM Chart

The Kanban Board is a workflow management tool that maximizes the potential throughput of work in process (WIP) by protecting the process from activity overload and the ensuing task-switching that is generated as a result. Task switching is also known as change-over time and is one form of waste.

The fastest way to deliver value through a system is to address one item at a time. The is sometimes called "Single-piece Flow." As items are completed one at a time, they can be deployed and start generating value sooner than later. Having work-in-process limits on the Kanban Board enables One-piece flow.

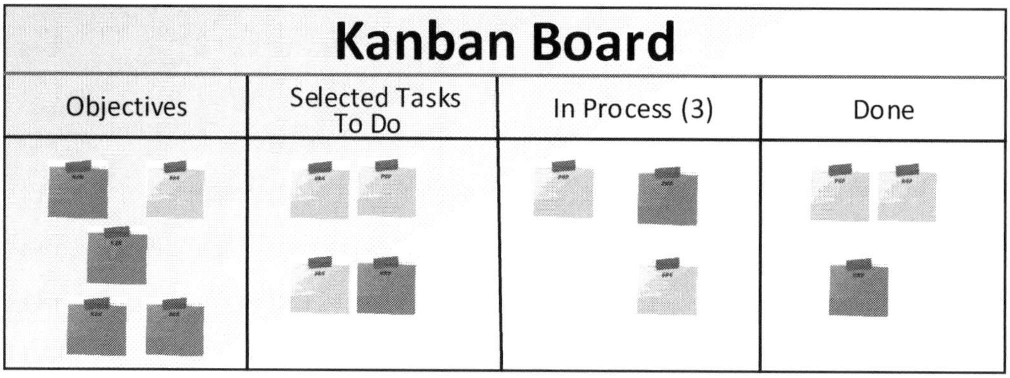

Kanban Board			
Objectives	Selected Tasks To Do	In Process (3)	Done

Figure 55 - Kanban Board

Also, as problems are experienced, Kata Cards are updated and in some cases Process Recovery Models and Asset Recovery Models are updated with root cause information and remediation steps so that future repeated problems can be recovered more quickly.

Notice the overarching principle at play here: A quality system empowers an organization so that a problem might occur once, but thanks to Kata Cards, and Recovery Models, the same problem should never impact the organization twice.

> A quality system empowers an organization so that a problem might occur once, but thanks to Kata Cards, and Recovery Models, the same problem should never impact the organization twice.

Chapter 17 - Using Stable in Implementation

Implementation is the deployment and training part of a product development project, which typically is the final set of the project's activities.

During Implementation, a tested and completed solution is deployed to a production site, typically at a customer location. Although each customer is unique, usually the same milestones must be reached to deploy a solution successfully.

A typical Implementation process contains at least these five steps: Site Discovery, Preparation, Initial Implementation, Full Implementation, and Sustainment.

Implementation Model

Figure 56 - Implementation Model

Implementation Using the Stable Framework™

Implementation activities are usually project-based. The project steps should be identified and processed through the Kanban. Each step or set of steps should be thought of as a repeatable process and should be identified, modeled into a process map, and have an SOP and a Kata Card created for it.

Implementation Model + Stable Framework™

- Discovery Kata Card
  - ☐ Infrastructure in place?

- Preparation Kata Card
  - ☐ Project Plan created?

- Initial Implementation Kata Card
  - ☐ Smoke Test successful?

- Full Implementation Kata Card
  - ☐ All sites up & running?
  - ☐ Training completed?

- Sustainment Kata Card
  - ☐ Help Desk info in place?
  - ☐ Customer billing in place?

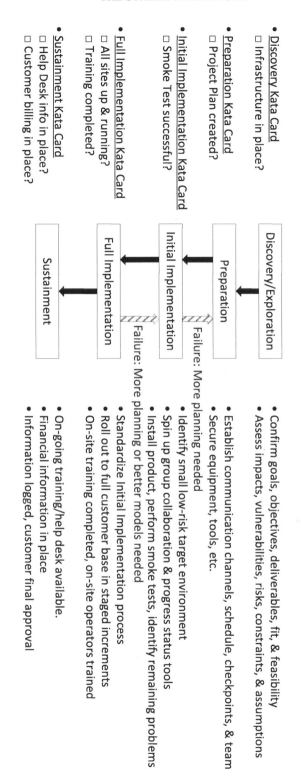

- Confirm goals, objectives, deliverables, fit, & feasibility
- Assess impacts, vulnerabilities, risks, constraints, & assumptions

- Establish communication channels, schedule, checkpoints, & team
- Secure equipment, tools, etc.

- Identify small low-risk target environment
- Spin up group collaboration & progress status tools
- Install product, perform smoke tests, identify remaining problems

- Standardize Initial Implementation process
- Roll out to full customer base in staged increments
- On-site training completed, on-site operators trained

- On-going training/help desk available.
- Financial information in place
- Information logged, customer final approval

Figure 57 - Implementation Model using The Stable Framework™

As the implementation Process Owner performs the steps in the process, the PRO submits a Kata Card showing that each item has been checked off and the process was completed successfully. Think of a completed Kata Card as a certificate authenticating that the work was completed successfully. The Kata Card is collected at the next checkpoint by the recipient PRO, or given to the Master Chief if it's the final step in the value stream.

Kata Cards for each step add value by identifying past problems that future implementers can be protected from.

For example, in the Discovery step, after sending someone out to a rural customer site who may still be on a slow T100 network connection when 1GB is required, the implementation team learns in the future to run a network speed test before all future site discovery visits. They would then add this step to their "Site Discovery" Kata Card so that future installs don't experience this problem.

For preparation, the team may have learned to review the IP addresses at the facility with the local IT Director and have their hardware preconfigured to save time when they arrive. This would be added as a step to the "Preparation for Install" Kata Card.

For the Initial Implementation step, the group would eventually learn to perform a complete smoke test on site after the system has been installed. A smoke test is a test designed to check a single feature from every screen or all major screens after installation to be certain everything is present and working without spending all the time required to do a full test at a location. This would be added to the "Initial Implementation" Kata Card.

As another example, after discovering that the less expensive WIFI repeaters really didn't work as planned, the installers can add a step on the Kata Card to use a WIFI signal strength meter and walk through all boundaries of intended use to prevent this problem from happening at all future install locations.

If the initial implementation went well, the team would then do a full deployment to all relevant workstations at the location using a "Full Implementation" Kata Card. Here they could also collect information such as the number of licenses deployed.

Finally, a "Sustainment" Kata Card could handle the remaining items such as training expectations, ongoing support payment methods, and so on.

Over time these repeated processes will benefit as more good ideas and precautions are identified and added to the Kata Card for each step. This reduces errors, rework, and mistakes, making the implementation process faster and less expensive over time.

Chapter 18 - Using Stable in DevOps

DevOps is the technical skill support group tasked with performing a critical part of the software value stream: deployment. Agile development techniques enabled teams to create releasable software more frequently, creating the need to coordinate multiple product releases, and in many cases dependent releases. The orchestration and progress of these activities is often called Continuous Integration/Continuous Delivery, or the CI/CD Pipeline.

In addition, if your organization is offering Software as a Service (SaaS) products, or Platform as a Service (PaaS) products, the need for Development to collaborate with Operations increased the need for excellence in DevOps.

According to Gene Kim, co-author of *The DevOps Handbook*, and *The Phoenix Project*, there are Three Ways that direct DevOps efforts:

1. The *First Way* emphasizes the performance of the entire system – the value stream.
2. The *Second Way* is about shortening and amplifying feedback loops.
3. The *Third Way* is about creating a culture that fosters continual learning and understanding.

The Stable Framework™ enables these Three Ways in DevOps, whether your team is Agile or Plan-based. Your model may differ, but in this model, coding is completed, the programmers have completed Test Driven Development (TDD) testing and confirmed the application works correctly with the end users before checking in the code.

Traditionally, User Acceptance Testing (UAT) is one of the final steps in this process, but because we are trying to streamline the time it takes the team to complete these steps in this model it makes sense for the end user to approve the final functionality before the code goes through all the downstream component testing, regression testing, and load testing steps. This is the second principle in the *Theory of Constraints: Exploit the Constraint.* In other words, why do all that downstream testing if you're not sure the user will accept it as designed? Obtain user approval before performing all that time-consuming, expensive work.

Once the code is checked in, the DevOps team updates the automated test scripts and any deployment scripts that will have changed with the new features. You may have a group of testers who component test and System Test (Regression Test) the whole build at this point.

DevOps Model

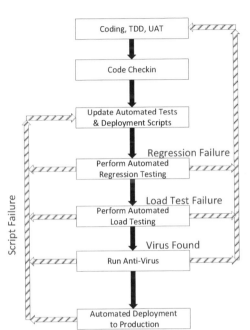

Coding, TDD, UAT	• Coding • Test Driven Development (TDD) • User Acceptance Testing (UAT) • Confirm updated Version Number
Code Checkin	• Continuous Integration • Automatic Build increment
Update Automated Tests & Deployment Scripts	• Minimal effort needed to adjust automated Install/Update, Regression, Load & Deployment Scripts
Perform Automated Regression Testing	Regression Failure • Using tools, automatically Install/Update & Regression Test candidate build
Perform Automated Load Testing	Load Test Failure • Using tools, Load & Stress Test candidate build
Run Anti-Virus	Virus Found • If needed, automatically Anti-Virus test candidate build
Automated Deployment to Production	• Automatically deploy build to production • Send out notifications

Script Failure

Figure 58 - DevOps Model

If automated regression testing is engineered correctly it is structured like a main starting point sliced into multiple automated test scripts that all start and end at the same location. This way they can be daisy-chained together and only the scripts impacted by the new requirements in the candidate release need to be created or adjusted. A well-trained group can recreate updated automation slices for the specific changed portions of the new build while testing it. This way they minimize the effort and time required to test what changed and then regression test the whole build.

Assuming no issues, they then load test and stress test the new build. Load testing confirms the build meets any established Service Level Agreements (SLA's), and Stress Testing gives the engineering team an idea of how far beyond the SLA the new release will function reasonably.

Finally, if everything passed successfully, they run the standard anti-virus check, and then the automated deployment scripts to push the production-ready build to its proper production destination.

If anything should fail along the way, a feedback loop exists to promptly alert an operator at a previous step that something failed, and adjustments to automation scripts or bug fixes would need to be made before the next attempt.

DevOps Using the Stable Framework™

Each value proposition a company offers has it's own unique set of release logistics. Therefore, a company may have one or more CI/CD Pipelines, depending on the number of products or services it offers.

Due to the many handoffs and the complexity of the last few steps in a value stream, the team can easily create a lot of waste by repeatedly making the same mistakes. Using the Stable Framework™, Process Owners would have tight feedback loops which lower the probability of making mistakes as they learn from their past, resulting in a faster and leaner stream of value to the customer.

With Stable, the processes would be defined for each step in the diagram below. Kata Cards would be created for each step: A "Development" Kata Card would be used for completing the code and performing the Architectural, TDD, and UAT Reviews.

Then, a "Continuous Integration" Kata Card, a "Regression Testing" Kata Card, and an "Automated Deployment" Kata Card would be created.

Many of the alerts and customer communications can be automated, as well, and can appear as checkbox items on the Kata Cards.

With deployment, often some things must be temporarily turned off, for the new build to be deployed. A Kara Card ensures they are turned back on afterwards, every time.

Process Recovery Models and Asset Recovery Models would be on hand to quickly bring failed systems back online with the guidance of lessons learned from past failures. This way, no steps get skipped, and everything happens with minimum error.

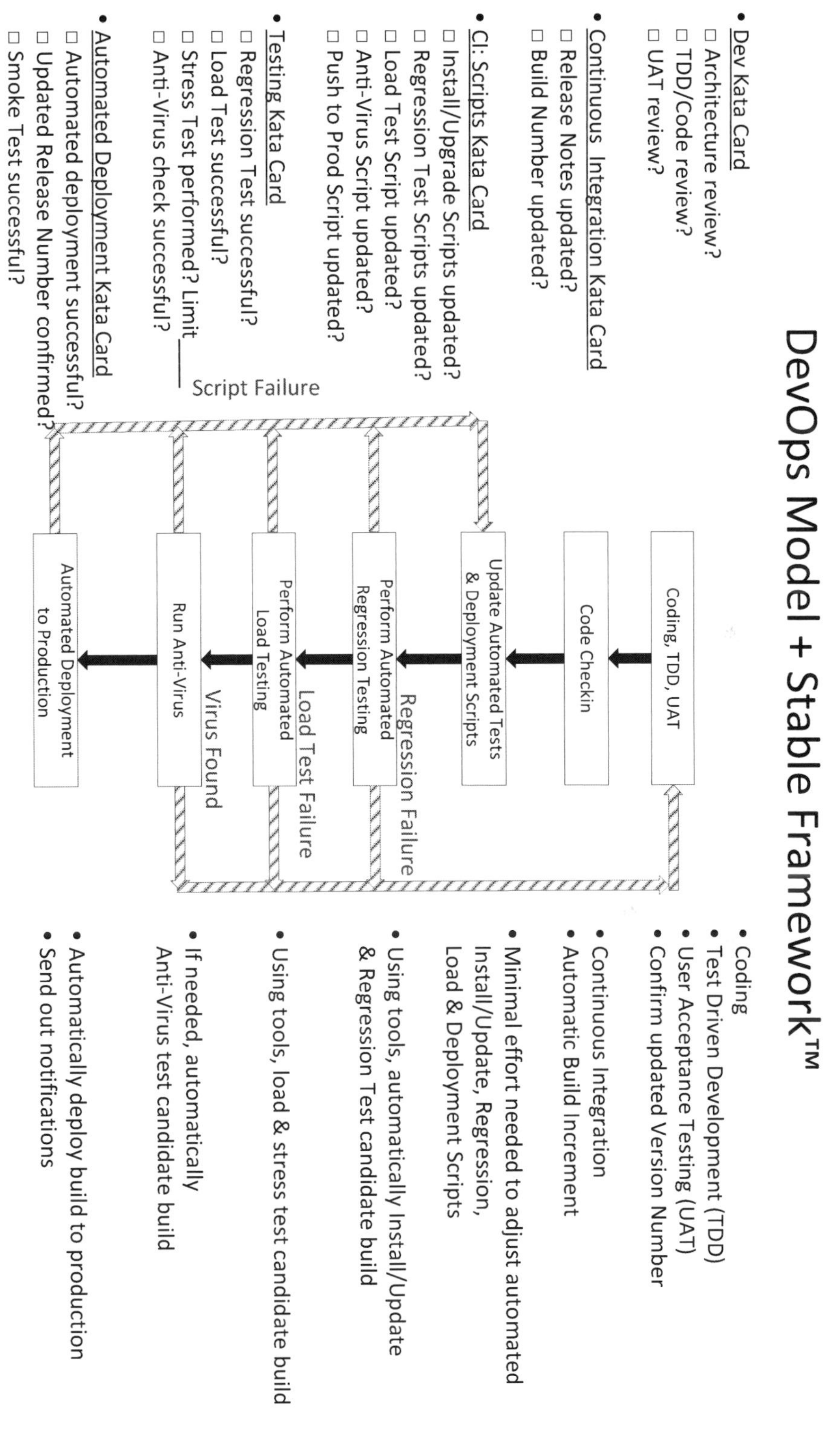

DevOps Model + Stable Framework™

- Dev Kata Card
  □ Architecture review?
  □ TDD/Code review?
  □ UAT review?

- Continuous Integration Kata Card
  □ Release Notes updated?
  □ Build Number updated?

- CI: Scripts Kata Card
  □ Install/Upgrade Scripts updated?
  □ Regression Test Scripts updated?
  □ Load Test Script updated?
  □ Anti-Virus Script updated?
  □ Push to Prod Script updated?

- Testing Kata Card
  □ Regression Test successful?
  □ Load Test successful?
  □ Stress Test performed? Limit ___
  □ Anti-Virus check successful?

- Automated Deployment Kata Card
  □ Automated deployment successful?
  □ Updated Release Number confirmed?
  □ Smoke Test successful?

Script Failure

Coding, TDD, UAT

Code Checkin

Update Automated Tests & Deployment Scripts

Perform Automated Regression Testing — Regression Failure

Perform Automated Load Testing — Load Test Failure

Run Anti-Virus — Virus Found

Automated Deployment to Production

- Coding
- Test Driven Development (TDD)
- User Acceptance Testing (UAT)
- Confirm updated Version Number

- Continuous Integration
- Automatic Build Increment

- Minimal effort needed to adjust automated Install/Update, Regression, Load & Deployment Scripts

- Using tools, automatically Install/Update & Regression Test candidate build

- Using tools, load & stress test candidate build

- If needed, automatically Anti-Virus test candidate build

- Automatically deploy build to production
- Send out notifications

Figure 59 - DevOps Using the Stable Framework™

In addition, any production change request can be handled using the Customer Request Queue and the Kanban Board. Requests can be time stamped when work begins, and them time stamped when they are completed.

Categorizing these requests and tracking the amount of time for their completion can provide customers with accurate estimates regarding how long they need to wait for their requests to be fulfilled.

Chapter 19 - Using Stable in Product Development

Let's look at a how a quality program can apply to a product development environment. A well-known approach to developing a new product or service is called the V-Model, because it looks like a V. It contains the initial development steps from concept to completion down the left side of the model, and the corresponding quality control steps up from the bottom to the top on the right side of the V.

The Product Development "V- Model"

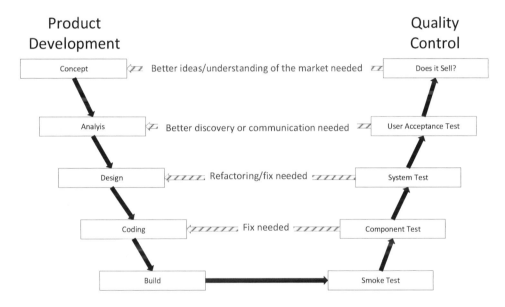

Figure 60 - Product Development V-Model

The development steps include the Initial Concept, Requirements Analysis, Architectural Design, and Coding.

The quality activities on the right represent Quality Control. These are the activities performed after a component has been built but before the customer receives it. You can think of these as inspection points.

The Smoke Test is important to make sure all the components have installed correctly. Component Testing is inspecting each part of a complex deliverable, such as individual forms, and reports. System

Testing is inspecting the entire deliverable, particularly data flow from one component to another within the whole system. User Acceptance Testing is showing the system to a customer and getting customer approval on the usefulness and accuracy of the new system compared to what was expected.

We call the inspections mentioned above Quality Control, because these inspections happen after the component has been built, but before the customer receives it. Quality Assurance, however, happens during the development of the component.

Product Development Using the Stable Framework™

Quality Control is product quality. It can be thought of as inspecting a candidate deliverable before the customer receives it.

Quality Assurance is process quality. It is your proof that a process step was followed correctly. Quality Assurance is the record of accountability that indicates each step was performed according to the best-known methodology available at that time. We call that methodology the Standard Operating Procedure (SOP). The SOP instructs the employee how to perform the pre-identified, repeatable task, and should include a checklist, called a Kata Card, of risks to watch out for.

This Kata Card is what gets signed, date-stamped, and handed to the receiving process owner when the work is complete. The actual SOP should be read by each practicing employee during initial onboarding, during periodic training, or when re-training is necessary. Retraining is necessary when a step is performed contrary to the SOP, or a completed Kata Card result proves to be inaccurate. This will become apparent when the root-cause of a nonconforming item was found to be the result of an employee not completing a Kata Card properly.

In product development, Quality Assurance activities include a Concept Review, Requirements Review, Design Review, and a Code Review. If you are using Stable outside of software, you can think of Code and Code Reviews as Construction and Construction Reviews.

A review works as follows. A Kata Card is created to govern that type of review. The Kata Card contains known risks and past problems found

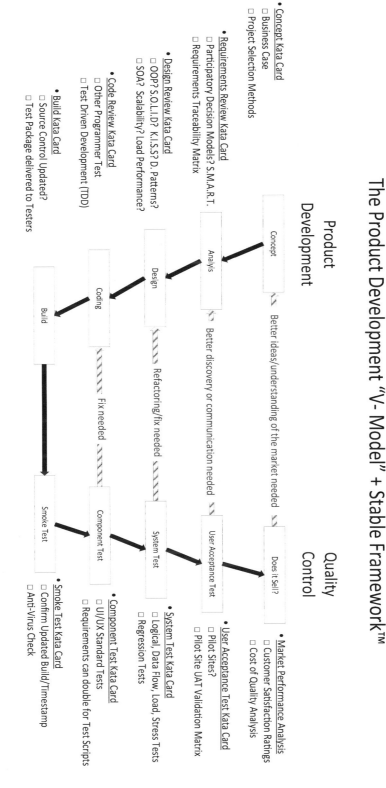

The Product Development "V- Model" + Stable Framework™

Product Development

- <u>Concept Kata Card</u>
  - ☐ Business Case
  - ☐ Project Selection Methods

- <u>Requirements Review Kata Card</u>
  - ☐ Participatory Decision Models? S.M.A.R.T.
  - ☐ SOA? Scalability? Load Performance?
  - ☐ Requirements Traceability Matrix

- <u>Design Review Kata Card</u>
  - ☐ OOP? S.O.L.I.D? K.I.S.S? D. Patterns?
  - ☐ Other Programmer Test
  - ☐ Test Driven Development (TDD)

- <u>Code Review Kata Card</u>
  - ☐ Source Control Updated?
  - ☐ Test Package delivered to Testers

- <u>Build Kata Card</u>

Concept → Analysis → Design → Coding → Build

Better ideas/understanding of the market needed

Better discovery or communication needed

Refactoring/fix needed

Fix needed

Quality Control

Does it Sell? → User Acceptance Test → System Test → Component Test → Smoke Test

- <u>Market Performance Analysis</u>
  - ☐ Customer Satisfaction Ratings
  - ☐ Cost of Quality Analysis

- <u>User Acceptance Test Kata Card</u>
  - ☐ Pilot Sites?
  - ☐ Pilot Site UAT Validation Matrix

- <u>System Test Kata Card</u>
  - ☐ Logical, Data Flow, Load, Stress Tests
  - ☐ Regression Tests

- <u>Component Test Kata Card</u>
  - ☐ UI/UX Standard Tests
  - ☐ Requirements can double for Test Scripts

- <u>Smoke Test Kata Card</u>
  - ☐ Confirm Updated Build/Timestamp
  - ☐ Anti-Virus Check

Figure 61 - Product Development V-Model Using the Stable Framework™

during that step. The reviewer can be a single person or a committee, and they go through the Kata Card with the author, indicate their name, date, and time, and submit the Kata Card as proof the review was conducted.

It is important that the completed Kata Card be collected by the Process Owner of the next step. If the next step is the end of the value stream, the final Kata Card and all preceding Kata Cards go to the Master Chief, who puts them in the Kata Card historical log. When the project is completed, the review records become part of the project historical documentation.

A Concept Review is the most elaborate review. We call this a Business Case Analysis and leave it to the strategists and portfolio managers to vet an idea and commission it to the design team.

A Requirements Review can be a small committee of Business Analysts and Quality Engineers who scrutinize the requirements for any areas of unreasonable ambiguity or past requirement mishaps.

In a situation where a new requirements risk is discovered, the risk should be added to the Requirements Review Kata Card so that future projects will benefit from that knowledge.

An architecture review works the same way. An "Architecture Review" Kata Card is created and added as the company learns about new risks. It may begin with commonly known architectural best practices and get expanded as future problems are experienced.

For example, while designing an enterprise software solution, it's important to source the date and time used by the software from the local database server. This way, all the workstations are synchronized and logging is consistent. Another design consideration might be to ensure the consistency of all Error Messages, Reports, Screens, and Screen Menus, so it looks like one person created the whole system instead of multiple people creating different parts of it.

In time, the group can add more items of design interest to their "Design Review" Kata Cards such as security items, and then whoever conducts the design review must inspect the design for each entry on the Kata Card, sign and date the "Design Review" Kata Card, and submit it to the next Process Owner downstream, repeating this process until the Master Chief receives the final stack of Kata Cards for that release and enters them into the permanent project history.

A code review can be conducted by another coder. The team would have a common "Code Review" Kata Card, which lists coding habits that are not permitted. For example, the code must have a healthy ratio of comments per lines of code. Or, the code must contain readable variable names, etc.

As time goes on, this Kata Card can be added to as new issues are found. Again, the reviewer records his or her name, the author's name and work, and time stamps the kata card, and then it becomes part of what is handed off to the owner of the next step in the process.

This is the basic work of Quality Assurance. It can be expanded to include a review of the right side of the V-Model. If testing is not producing good results, each QC task can be QA'd. A "Component Test" Kata Card can be created and used for a Component Test Review. An "Integration Test" Kata Card can be created for each Integration Test. A "System Test" Kata Card can be created for conducting System Test Reviews, and a "User Acceptance Test" Kata Card can be created for conducting a User Acceptance Review. As a company experiences natural problems and shortcomings, it can add the root causes as items to check for on the respective Kata Cards to ensure that the past problems never happen again. This is what it means to be a learning organization.

For some projects, you may even consider adding a decision log to the project, where critical decisions are recorded, timestamped, and key information is noted about them such as who came up with the idea, and who agreed to make the decision.

Chapter 20 - Using Stable with Agile Development

Agile development originated in the software industry and is an excellent model for delivering value quickly when there are a lot of unknowns in the project environment.

Scrum, the most popular implementation of Agile, was invented in 1993 by Ken Schwaber and Jeff Sutherland.

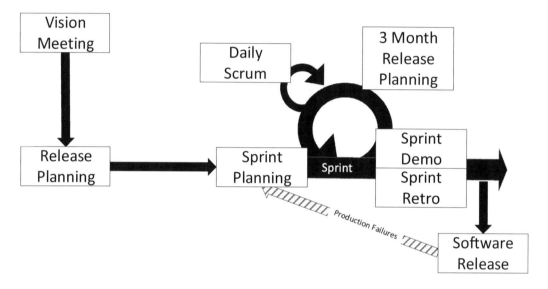

Figure 62 - PMI-ACP® Scrum Model

Many implementations of Scrum exist today. I'm displaying the PMI-ACP® Scrum model because I think it is a well-developed and useful implementation. It is the core Scrum Framework with some added accessories that enable it to scale well for large projects.

The PMI-ACP® model provides a Vision Meeting to charter the project, which means establishing the project objectives and success criteria, clarifying what the project should produce, who the intended customers are, and why the project is being executed.

It also includes a Release Planning Meeting where large projects can have analysis, design, and proper estimation performed for them, giving the stakeholders a reasonable idea of how long the project will take, and how much the project will cost.

PMI-ACP® Scrum Model + Stable Framework™

- Vision Meeting Kata Card
  - ☐ Sponsor present?
  - ☐ End User present?
  - ☐ Team present?
  - ☐ Problem Statement discussed?
  - ☐ Objectives, Risks, Constraints?

- Release Planning Kata Card
  - ☐ Real User Input?
  - ☐ MVP/MMF?
  - ☐ Scalability?
  - ☐ Security?
  - ☐ SOA?

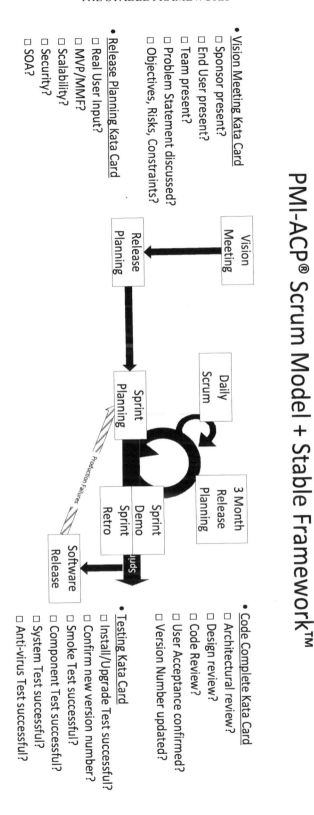

- Code Complete Kata Card
  - ☐ Architectural review?
  - ☐ Design review?
  - ☐ Code Review?
  - ☐ User Acceptance confirmed?
  - ☐ Version Number updated?

- Testing Kata Card
  - ☐ Install/Upgrade Test successful?
  - ☐ Confirm new version number?
  - ☐ Smoke Test successful?
  - ☐ Component Test successful?
  - ☐ System Test successful?
  - ☐ Anti-virus Test successful?

Figure 63 - PMI-ACP® Scrum Model Using the Stable Framework™

Agile Product Development Using the Stable Framework™

In the Vision Meeting, where the project is chartered, a "Chartering" Kata Card can ensure all common objectives, constraints, risks, industry regulatory restrictions, competitive products, and end-users have been considered.

In the Release Planning Meeting, a "Design" Kata Card can ensure that proper design patterns, scalability concerns, security issues, and any Service Oriented Architecture needs are met.

For development, "Code Review," "User Experience" and "User Acceptance" Kata Cards should be completed before a developer can integrate any code.

All the "Quality Assurance" Kata Cards apply to the testers on the team. "Smoke Test" Kata Cards, "Component Test" Kata Cards, "System Test" Kata Cards, "Interface Test" Kata Cars, "Load Test" Kata Cards and finally "User Acceptance Test" Kata Cards can be used to be sure the tests are capturing everything desired. Perhaps they can be combined into one document.

Chapter 21 - Stable Portfolio Management

Portfolio Management is a critical component for an organization that wants to stay competitive. As companies grow, they want to apply their resources to those investments that are in their best interests. Portfolio Management is about choosing just those project opportunities that best serve the objectives of the organization, and saying no to the rest.

Portfolio Management

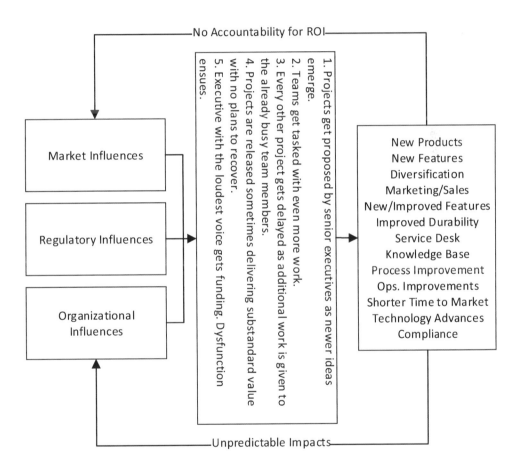

Figure 64 - Portfolio Management Model

Most organizations have no defined Portfolio Management process, and many have no idea what the term even means. In scenarios like this, projects simply get recommended by senior executives in various business areas. Little care is taken to throttle the workload placed on the teams, causing overburdens leading to frustration, thrashing, and delays everywhere. Post project analyses are rarely performed, leaving lots of lost value and revenue on the table for competitors. Poor portfolio visibility leads to a constant imbalance of priorities for the company making their written strategy little more than talk.

Portfolio Management using The Stable Framework™

An organization is a value system, complete with inputs, outputs, and throughputs. The outputs of the system are the value propositions that attract customers. We call that Customer Value. The inputs to the system are market forces, and various forms of capital. We call that Opportunity. The sustainment and growth a company experiences as a result of exploiting opportunities to generate customer value, we call Business Value. Whatever will get an organization to where it wants to be next has high business value. Business Value is best expressed using Key Performance Indicators (KPI's).

Think of Portfolio Management as a mechanism to direct your organization's value system. Specific inputs to this system are market reports, product and service performance summaries, competitive analysis, regulatory requirements, and organizational influences. Outputs are product and service performance, market share growth rates, and internal quarterly business assessments. Throughput factors within the system are leadership, culture, skills, and knowledge proficiency. In other words, how well the organization executes.

A good Portfolio Management process ensures the following:

- Corporate strategy is defined and clear.
- Market influences and competitor trends are monitored.
- Project teams are not overburdened.
- Projects are prioritized to reflect the current corporate strategy.
- High and low risk is balanced during project selection, as needed.
- Low value projects get cancelled when necessary.
- Communication is improved between business and IT.
- Project value yield is monitored for some time after completion.
- Transparency and feedback are improved as all major project efforts are visible to the executive team.

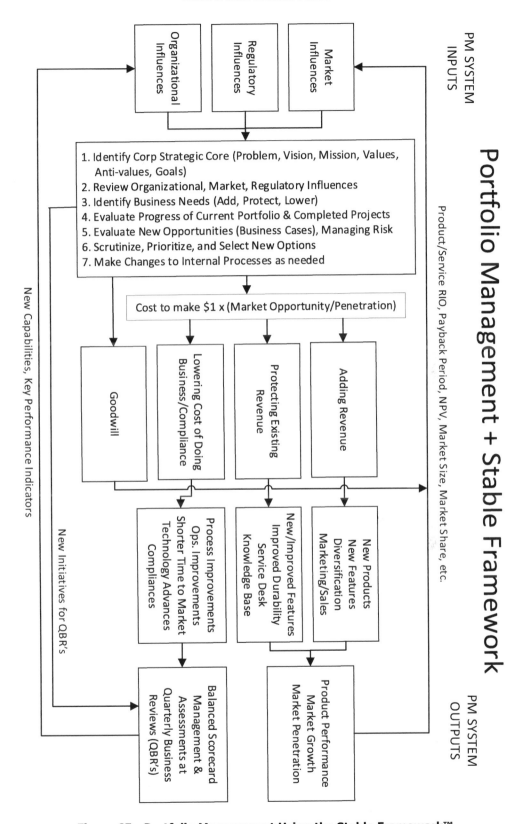

Figure 65 - Portfolio Management Using the Stable Framework™

Typically, a strategy review happens annually. Quarterly Business Reviews are conducted to assess market, product and service performance. And the Portfolio Management Committee meets every two weeks to prioritize project approvals based on the latest information and opportunities at hand.

The committee is comprised of the senior executives within the organization, and typically the VP's and Directors of all of the functional units.

Following is a description of a basic Portfolio Management process:

1. Identify the organizations Strategic Core (Vision, Mission, Values, Goals)

> The Strategy Review meeting happens annually, or sometimes every six months. This is where the CEO and board of directors and senior executives decide the corporate strategy for the organization going forward. It's important to understand that strategy is not simply what the group decides, but rather how they spend the organizations money towards that decision. In a well-run portfolio management process, the majority of the projects that get funded reflect the decided upon strategy.

> As part of the strategy review, if the following core is not already in place, the executive committee should clarify the Corporate Vison, Mission, and Values.

> A Vision Statement is a description of a future state, which directs the company's overall purpose of existing. A Mission Statement indicates how they intend to achieve the corporate vision. Value statements help guide decision making and prioritization.

> Once these are in place, the organization can set its present strategy, which means its long-term goals. Long term goals drive large strategic initiatives, and should reflect the Mission Statement and Values of the organization.

> Short term goals are subsets of long-term goals and drive department scorecards which get reported on quarterly during the Quarterly Business Review (QBR) meetings.

> Be sure everyone participating in the Portfolio Management meeting is aware of and clear about the organizations current Vision,

Mission, Values, and Goals. It may be appropriate to create a performance console on the wall in the Portfolio Management meeting room, and have all of this information posted.

2. Review Organizational, Market, and Regulatory Influences

Quarterly it is beneficial to have marketing prepare a report of marketplace trends, competitor trends, and any regulatory changes that are pending or have emerged. It's important to be aware of what your largest or fastest moving competitors are doing.

In addition, it's important to include trending keywords in your products and services where applicable, to stay relevant in your marketplace.

3. Evaluate Progress of Current Portfolio & Completed Projects

This step requires a Product Performance Inventory, and Master Project Schedule.

Product Performance Inventory

The Product Performance Inventory is a list of all of the primary products and services currently offered by your organization, along with information about each one such as market size, market share and annual revenue. It may be helpful to include a column showing

		Estimated				Net
Name	Annual Revenue	Market Size	Market Share	Competitive Pressure	Customer Satisfaction	Promotor Score
Product 1	$2,600,000	$20M +	13%	4	9	8
Product 2	$400,000	$20M +	2%	6	7	6
Product 3	$25,000	$5M +	0.50%	8	6	4

Figure 66 - Product Performance Inventory

the % change since this time last year. This list should match the public entries on the Operations Cumulative System Performance Chart, and be owned by Marketing and populated by marketing research. If revenue is kept private, a revenue rank can be substituted.

An array of decision support tools exists to examine the market performance of products or services, such as the Boston Matrix, Bowman's Strategy Clock, and Porter's Three Strategies. Your product managers or marketing team should provide up-to-date reports of these products to be used for analysis.

In addition, recently completed projects or services should have a post-release ROI evaluation from the project sponsor, or an assigned business analyst. These post-delivery ROI evaluations are typically performed quarterly until trends emerge. Information collected here is critical. Sometimes new products fail to meet their intended ROI and with just a small adjustment can become high-demand products. Consequently, these evaluations sometimes trigger high-yielding follow-on projects.

Master Project Schedule

The Master Project Schedule is a list of all current projects being worked from all departments, their status, timeline, and costs to date. In some cases, costs extracted from this information gets billed

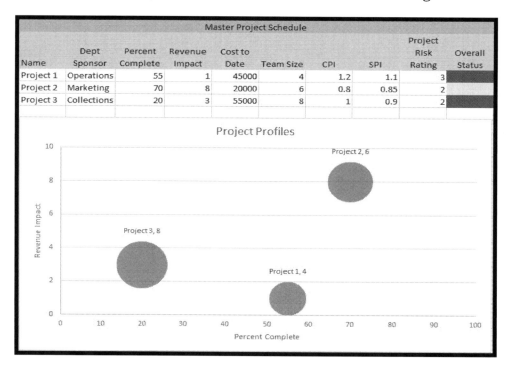

Name	Dept Sponsor	Percent Complete	Revenue Impact	Cost to Date	Team Size	CPI	SPI	Project Risk Rating	Overall Status
Project 1	Operations	55	1	45000	4	1.2	1.1	3	
Project 2	Marketing	70	8	20000	6	0.8	0.85	2	
Project 3	Collections	20	3	55000	8	1	0.9	2	

Figure 67 - Master Project Schedule

back to internal departments who requested the projects from IT so that internal spending can be tracked for the organization.

The example above shows percent complete, expected revenue impact, and team size. Any metrics of interest can be tracked and reported on using a similar chart.

Projects that are failing, or candidates for cancellation can be examined and discussed with this tool, to aid in decision support.

4. Prioritize Project Opportunities (Business Value)

This step is different depending on the type of approach your organization takes to determine how it wants to organize choices.

One common set of categories to organize projects into is:

1. Compliance
2. Protect Existing Revenue (Improvement requests, robustness)
3. Add New Revenue (New products)
4. Lower the Cost of Doing Business (Efficiency)
5. Goodwill (Optional, to improve community image)

For decades, companies have debated which order is the best to use for prioritizing projects, and the debate continues today. In Stable, we recommend the Sandcone Model approach, which is the priority listed above. It was used by Kia/Hyundai to outperform Toyota over time.

Before we discuss the approach, a little portfolio theory helps. Business owners need to understand that one of the first critical investments needed, after creating their first great product or service is a marketing and sales engine. A second or third great product will not make a first great product sell if the organization has no marketing and sales engine. One of the first priorities for a company who wants to grow is to invest in a marketing and sales engine.

Secondly, efficiencies are more critical for small companies than large companies. If you have a small company, you may want to switch number 3 and 4.

Third, understand that all companies want to add new revenue, but it's cheaper to protect existing revenue once you have a sellable

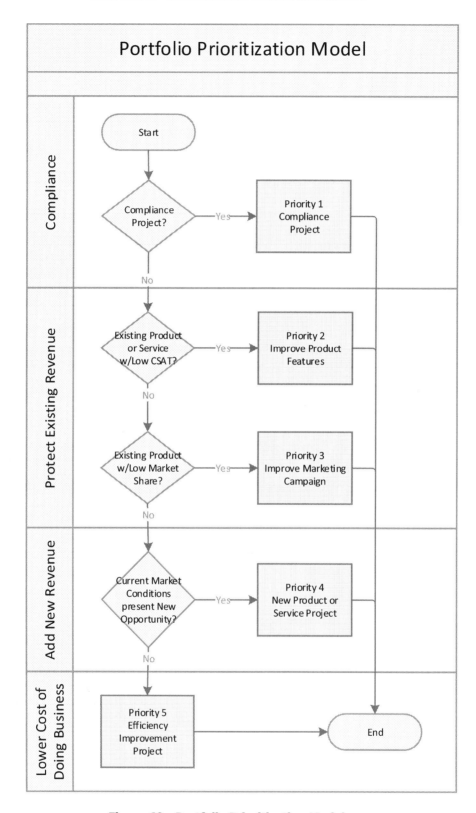

Figure 68 - Portfolio Prioritization Model

asset, than it is to develop a whole new one. If you have an asset that needs a little help to improve its customer satisfaction scores, or market penetration rates, your money is better spent there than on a brand-new product.

Also, the current market share of a good product is an important consideration. If you have a product with healthy customer satisfaction (CSAT) scores and a low market share (0-16%), you are better off investing in a marketing campaign for that product then spending money developing anyting new. See if you can get it into the early majority market share range (17%+) where it will begin to sell itself through word-of-mouth.

If your existing products have good CSAT scores, and good market share, it's time to examine the market for new opportunities. If some exist, it may be time to build a new product or service. If no good new ideas exist in your marketplace, you might consider diversifying existing products into alternate markets, or alternate price points.

If none of these options seem attractive, it may be a great time to fund a project that lowers the cost of doing business, or produces goodwill in the community, while waiting for market opportunities to emerge later. The figure on the preceding page provides a logical flow for prioritizing opportunities using the Sandcone Model.

Be aware that a corporate perspective is not the same as the customer perspective. An investment promising a great ROI from the business perspective might hold little value for the customer base. Instead, the customers may be in dire need of a different product with a lower ROI. Always remember your customers operate inside of systems and you need to understand them and their system of needs to provide complete solutions for them.

5. Evaluate Large Opportunities (Business Cases) while Managing Risk

Any proposal costing over, for example $25,000, should be submitted with a Business Case. Smaller projects can fit in between the larger projects.

A Business Case is a persuasive document prepared by a project sponsor or their business analyst to explain why the proposed project would be a good investment for the company to pursue.

Business Cases contain basic components such as:

1. Executive summary of the concept.
2. S.W.O.T. Analysis (Strengths, Weaknesses, Opportunities, Threats within the current scenario).
3. Expected ROI (Can included multiple scenarios such as minimal/medium/maximum investment).
4. Expected payback period.
5. How the project benefits will be measured, and who will measure them.
6. Risks, constraints, anticipated budget, anticipated timeline.
7. Anticipated team size.
8. A getting started plan.

The sponsoring executive presents the Business Case, or lets the invited sponsor present the Business Case to the Portfolio Management team to be scrutinized.

6. Scrutinize, Prioritize, and Select the Next Project

As business cases are presented, the committee discusses them and identifies weaknesses in each proposal. If a substantial amount of weaknesses is identified, the sponsor is challenged to return with an improved proposal at a later meeting.

Figure 69 - Opportunity Profile Radar Chart

The group may want to establish a set of criteria critical to success, and then profile the top opportunities using a radar chart. A prioritization matrix can also be used to decide the most desirable project funding order.

Simple Project Selection Matrix							
	Project 1	Project 2	Project 3	Project 4	Project 5	Project 6	Project Value
Project 1	0	1	1	1	1	1	5
Project 2	0	0	0	1	1	0	2
Project 3	0	1	0	1	1	0	3
Project 4	0	0	0	0	1	0	1
Project 5	0	0	0	0	0	0	0
Project 6	0	1	1	1	1	0	4
The projects with the greatest Project Value are rated highest.							

Figure 70 - Project Prioritization Matrix

The Engineering Director and Project Director should be present in the meeting to identify the teams becoming available to begin work on the highest prioritized projects.

Too much task switching within a team slows down all the work being performed by the team by creating artificial delays. It's best to protect the team's bandwidth so they can expedite value faster into the marketplace. Remember, we want One-piece flow.

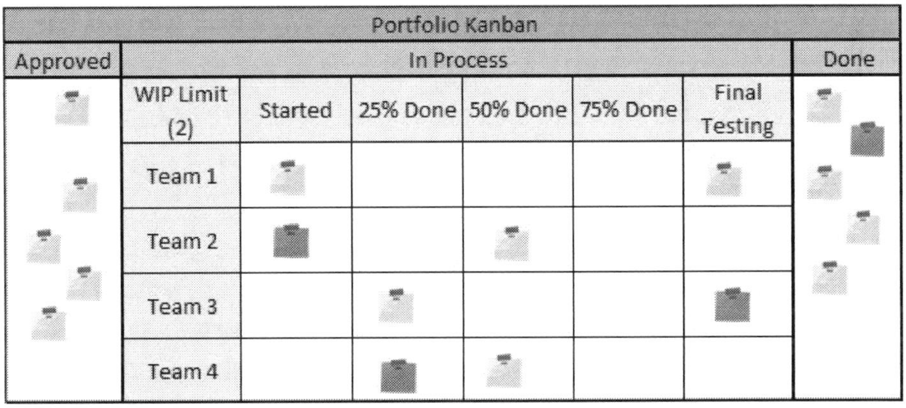

Portfolio Kanban							
Approved	In Process						Done
	WIP Limit (2)	Started	25% Done	50% Done	75% Done	Final Testing	
	Team 1						
	Team 2						
	Team 3						
	Team 4						

Figure 71 - Portfolio Kanban

Some groups list ongoing projects on a Portfolio Kanban Board to protect Work In Process limits established by each project team's natural bandwidth limitations.

The Portfolio Kanban Board might govern all projects started since the newest strategy change happened, or it might track work from year to year.

To maintain a consistent flow of project effort over time, teams that work well together are encouraged to stay working together from one project to the next. These teams are considered corporate assets and competitive advantages.

We've learned through research published by the International Software Benchmarking Standards Group (ISBSG.org) that the optimum team size is four people, with the ceiling being seven. Try to keep your core team sizes split in increments of four, and no larger than seven.

7. Make Changes to Internal Processes as needed.

A final output of a Portfolio Management meeting might include some takeaways for department managers to add to or adjust their quarterly business review goals. If a company uses a Balanced Scorecard, or some similar mechanism to report progress within their departments quarterly, they may be able to use takeaways from this meeting for that purpose.

Using Stable Framework tools and process controls will ensure all necessary considerations are made when needed.

A "Portfolio Management" Kata Card would include these checks:

- Is corporate strategy up-to-date, and visible?
- Have market trends, competitor trends, and regulatory trends been reported on within the past quarter? Are the findings visible?
- Have current status of products & services been reviewed?
- Have current status of projects in process been reviewed? Can any project benefit strategically from more resources?
- Has business opportunity driver been identified (Add, Protect, Lower Cost, Goodwill)?
- Have new proposals been presented?
- Have new projects been approved and prioritized with group input?

A "Market Update" Kata Card might include these checks for discussion:

- Market description.
- Market size (potential number of customers in terms of units sold, or total dollar value)?
- Current market trends?
- Emerging market trends?
- Largest market players and partnerships?
- Regulatory trends?
- Market vocabulary?

A "Product Analysis" Kata Card might include these checks for reporting:

- Market penetration rate?
- Competitive products?
- Competitive trends?
- Customer satisfaction trends?
- Net Promoter Score trends?
- ROI?

These are initial suggestions for Kata Card contents. As lessons are learned about markets, user experiences, regulatory influences, and other factors, checks can be added to appropriate Kata Cards to help the organization make better decisions going forward.

Post-project RIO evaluations will be a valuable source of updates to these Kata Cards. In highly competitive markets, experience gained in these areas sometimes is the only competitive advantage that a leading company has.

Overall, Portfolio Management can move a company towards their goals faster by maximizing their spending towards their corporate objectives, maximizing throughput by protecting work in process limits, and providing transparency between all parties on how business and IT are connected. If done properly the company will steadily grow with appropriate well-placed decisions.

Chapter 22 - Getting Started with Stable

Organizational transformations are challenging. In order to successfully transform your team, department, or entire organization to use a quality framework like Stable some things need to be in place.

The primary factor for success is executive support. From the top of the company down to the employees working the process steps, there needs to be a visible endorsement of the quality framework adoption program within the company.

Second, your employees need proper training. The best way to get it right from the beginning is to pursue training for your group. Training provides consistent messaging and expectations for each team member and has wonderful side effects such as improved team cohesion and an enhanced sense of employee self-worth by seeing the organization invest in each employee.

Third, your employees need to see visible progress and the positive effects of the new framework. The best way to do this is to identify metrics you can track before the transformation and then compare them to results after the transformation has been implemented.

Be sure to log the victories, post them in a public place, and share them with your teams and senior management.

You may also consider hiring a Stable Coach temporarily to assist with making sure the process has been implemented correctly, and there are no bad habits forming. Think of this as insurance for your training investment.

The Path to Stability

So, how do we turn a chaotic environment into a well-managed world-class organization?

First, we need to have some common understandings. A system is a collection of one or more processes that are related to each other and pass along assets from various entry points to an exit point.

The proper way to engineer a system is from the end to the beginning, meaning, you first identify what the customer wants, then you backtrack to determine how you will make that possible.

As we build a system to govern the standard work efforts, the system will, in turn, help us execute our work better. Better, in this case, means with fewer errors, and executed correctly the first time. As mistakes happen, the culture of the environment should be that the Master Chief and Process Owners ask themselves "How can we improve the system so that this can't happen again." In effect, we focus on improving the system instead of blaming the worker. This is a fundamental principle W. Edwards Deming taught. He told management it was their job to get the teams to improve the systems they work with so that their people couldn't repeat mistakes.

In order to understand a system, it's best to get outside of the system and analyze it from the outside looking in. Fish don't know they are in water. It's difficult for us to appreciate we are on an orb in space rotating at 1037 miles per hour.

We first must establish a system before we can improve it. We call this systemization, and it involves establishing a value proposition, a system flow, Standard Operating Procedures, and something called a Kata Card, which is a super-checklist. The Kata Card is where we add risk-averse tips learned from past experience. In the beginning Kata Cards may start out small, but they will grow over time. Kata Cards can serve as a platform for collecting data about the work. At the end of each process flow the Kata Cards are signed, dated, and given to the next downstream Process Owner, or the Master Chief if the Kata Card is the final step of the value stream.

A system can be slowed down for several reasons. Bottlenecks slow a system down. Ambiguous inputs slow a system down. Overburden can slow a system down. Inaccurate or incomplete operations due to incomplete training will slow a system down. Scrap and rework will slow a system down. Inspections and warranty work will slow a system down.

A system has only a certain amount of work it can handle at any one time, and a certain amount of work happening within it. We call the amount of work being done Work-in-Process (WIP) and the limit we call capacity or WIP-Limit. The closer we can get to one item being completed at a time the better. We call this "One Piece Flow." This concept enables us to put items of value into production much faster than waiting for a batch of items to be completed. For this reason, task-switching is always discouraged.

If you examine a system from end to end, somewhere within the system you will find the slowest moving step. This is the system bottleneck. The Theory-of-Constraints, an idea put forth by Dr. Eli Goldratt in a book called *The Goal*, teaches us it is important to find the slowest system bottleneck and improve its throughput, and then find the next slowest bottleneck and repeat the process until you are producing a product or service at the rate your customer consumes it. The rate at which your customer consumes your product is called Takt Time.

The Technology Adoption Lifecycle.

As new technologies emerge and are adopted by the masses, a common adoption pattern can be observed. We call this the Technology Adoption Lifecycle, and it is based on an older model originally called the Diffusion Process, written about by Bohlen, Beal, and Rogers in 1957. The model states that there are five types of customers when presenting a new idea: Innovators, Early Adopters, Early Majority, Late Majority, and Laggards.

The Technology Adoption Lifecycle

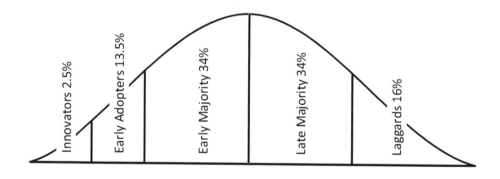

Figure 72 - The Technology Adoption Lifecycle

Innovators are the geeks who like to try new things just because they are intrigued. These people generally operate in high risk-tolerant environments, which means they usually haven't reached the point in their careers where they command budgets.

Early Adopters are people with budgets but are somewhat risk-averse. They will patiently wait for the innovators to experiment with the new technology before they try it out. They are interested in case-studies and documented proof of success.

The Early Majority is the first half of the market segment that adopts a rising star technology offered in its marketplace.

The Late Majority is the second half of the market segment that adopts a rising star offered in its marketplace. The Late Majority is usually distracted by budget needs elsewhere, older technology or policy limitations and cannot move as fast as the early adopters.

The Laggards are those who see the new technology as a benefit, but also a hassle. They are the last of the market to conform because they are complacent and happy with their current situations and will only change if forced by company policy, market demands, crisis, or technological isolation.

The Value Realization Curves

Another transformational dynamic to be aware of is the Time/Effort Value Realization curve, which demonstrates that it takes some

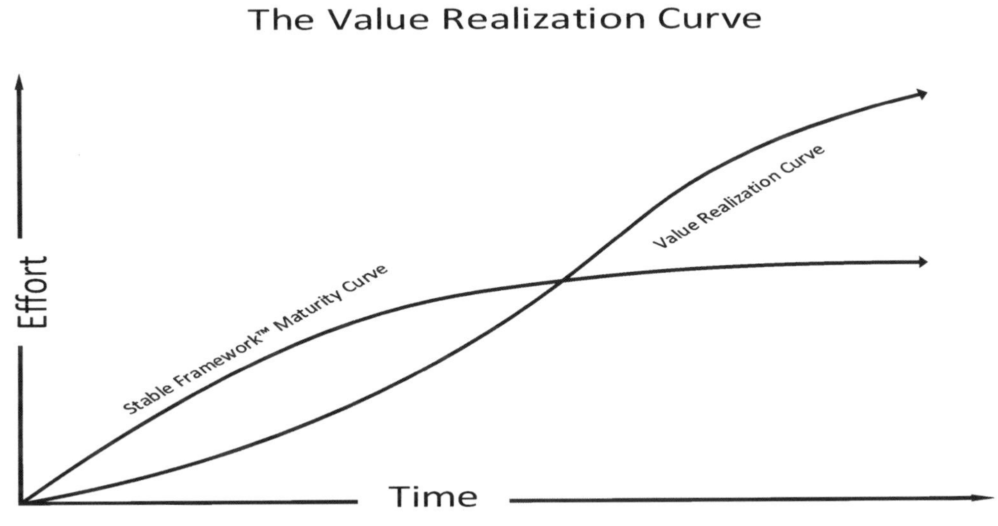

The Value Realization Curve

Figure 73 - The Value Realization Curve

consistent effort over time at first before progressive results start to emerge. For this reason, it's best to measure your environment before and after adoption, and to post victories publicly anytime your quality system saves anyone from rework, scrap, or any other problem.

Quality Planning

The initial implementation of Stable is called Quality Planning and is directed by the Master Chief. In this stage, the core structure and tools need to be made ready for the team.

The team should not be encumbered by having to figure out where to store process controls or where the standard Kata Card history is located.

Of course, the Master Chief can consult with the team on all these decisions, but in the end, if these items are not set up and in place, the Master Chief has not prepared the team adequately.

Quality Planning should be revisited every six months or so to be kept current.

The next section details the most common approach to setting up the Stable Framework™.

Initial Setup

1. Review The Stable Framework™ – Review the Stable Framework with your team. Discuss the challenges in your areas and the need for improvement. Get their commitment to learn the framework and agree on a go-forward plan.

2. Create a Configuration Management System – Identify or create a physical location where all the lists, templates, and other process controls will be stored and updated. Ideally, this should be a version-controlled document repository. If that is not available, use a shared network folder with appropriately named sub-folders that gets backed-up regularly.

3. Create a <u>System Schedule</u> - A schedule or calendar showing all the future planned events, including scheduled asset maintenance and any repeating timed events like end-of-month activities, or quarterly reviews.

2. Create a <u>System Backlog</u> - The System Backlog is a single location organized by four work queues: The Customer Request Queue, the Scheduled Activity Queue, the Equipment Maintenance Queue, and the CCAPA Queue. These separate queues are the landing site for incoming work. These can be as simple as lists in a spreadsheet or sticky notes on a wall.

3. Create a <u>Performance Console</u> - Otherwise known as an Information Radiator, this is a physical area at the work-site where queues, the Kanban board, work in process, and reporting assets are posted. In Lean, this concept is called Visible Management and we want to make all work identifiable and posted here.

4. Create a <u>Kanban Board</u> - A Kanban board will be the centerpiece on the Performance Console. This is a multi-column tool used to track the following work items: Objectives, Selected Tasks To Do, Work In Process, and Work Completed. Optionally, columns can be added for additional categories such as Pending Approval, Approved, On Hold, To Test, To UAT (User Acceptance Testing), etc. Your environment dictates which columns would be most appropriate to construct for your team.

5. Create <u>Historical Logs</u> - These logs could include a Production Change log, Kata Card repository, Training logs. and a SLAM Chart repository.

6. Create <u>Customer Satisfaction Surveys and Histograms</u> - Customer Satisfaction should always be measured using a histogram, showing changes over time. While the Process Owner is responsible for gathering customer information to improve the process outcomes, the Master Chief is responsible for gathering Customer Satisfaction data.

7. Create <u>SLAM Charts</u> - SLAM (Service Level Attainment Monitoring) Charts are the primary means for reporting on the success of a service over time. All primary services should have SLAM Charts posted for them on the Performance Console and should be updated daily. SLAM Histograms should also be tracked to show trends over

time. SLAM Charts are created and owned by the Process Owners.

8. Create a <u>Cumulative System Performance Chart</u> – Create a CSP Chart, and as it gets populated by performance information during the quarter, be prepared to present it at a Quarterly Business Review meeting, if your organization holds one.

9. Systematize <u>Value Streams</u> – Working with the team, identify all value streams within your area. Systematize each one by creating process maps, collecting customer information, expected Service Levels, and creating SOPs and Kata Card templates. Create Process and Asset Recovery Models, and Lessons Learned documents.

10. Determine <u>Length of Master Cycle</u> – Working with the team, identify the desired length of the Master Cycle within your areas. Anywhere from one to four weeks is good. If you work alongside an Agile team, it's best to match the Master Cycle to their Sprint pattern so that you can report together.

11. Establish <u>Master Cycle Meetings</u> – Establish the Master Cycle Planning Meeting, the daily Kaizen Stand-up Meeting time, and the Master Cycle Review Meeting, and Retrospective Meetings.

12. Establish <u>Inter-department Meetings</u> – If they don't already exist, you will need to establish a Change Control Committee meeting, a regular CCAPA meeting with the Service Desk, and be part of the engineering status update meetings.

13. Establish an <u>Alert System</u> – Off the shelf products exist for this purpose, or you can build your own. Start simple and add to it as needed. In time, it will become valuable.

14. Set up a <u>Training Log</u> – Establish a training registry showing who has been trained for each recurring process. Indicate the date/time so retraining can happen in six months or so.

15. Set up a <u>Suggestion Box</u> – Put a visible suggestion box near or on the Performance Console. Tell your group about it. Encourage suggestions and act on them.

Then, get started....

Be sure all your people get proper training. We learned from the Agile revolution that an untrained staff is far less effective than a properly trained staff.

Taking Over as Master Chief

If you assume the role of Master Chief in an existing area, these are the steps you should follow:

1. Confirm the location of the Configuration Management System.

2. Bring all Lists up-to-date.

3. Encourage all Process Owners to retrain and bring all their Standard Operating Procedures and Standard Kata Card Templates up to date, and to be sure the System Schedule is up to date.

5. Be sure the System Schedule is up to date by meeting with your executive sponsors and primary customers and recording anything relevant that your team will be involved in.

4. Be sure you have a Performance Console and it is kept up to date.

Sustaining the Stable Framework™

The key to a successful Stable Implementation is the transformation of everyone involved in the experience. Senior Management ought to understand that their support of the practices enables them to see at a glance the total and complete status of everything happening within the department clearly, leaving them free to spend their time growing their business. This is what is meant by Operational Excellence.

To the Master Chief, the Stable Framework™ means structure and guidance in introducing and applying techniques that empower teams and enable accountability at the same time. It's also a structured approach to transform tribal knowledge into institutional knowledge, and continually improve the most important metrics and the people within the Master Chief's department.

For the Process Owner, it provides structure, tools, and a realization that an invisible process exists which will protect them from future mistakes if they will respect it and work with it. The Stable Framework™ also provides the assurance that mistakes will be blamed on less-effective systems needing improvement, and not the operators.

For the customer, Stable provides metrics, visibility, and as much communication as the customer wants. If managed correctly, your customers' satisfaction levels should rise consistently over time.

For the supplier, consistent feedback from the Stable organization will provide clearer direction and insight into making the suppliers' services and products a better fit for its markets over time.

Chapter 23 - Data Collection & Analysis Tools

This chapter is devoted to showcasing a set of quality charts and tools useful for data collection and data analysis. They can be applied to Quality Planning, Initial Quality Assessment, Quality Assurance, Quality Control, and Continuous Improvement.

Data Collection Tools

Data Collection Tools are used to interpolate real world data into an organized form for knowledge generation, reporting or further analysis.

Context Diagram

The Context Diagram is useful for expressing the boundaries of a system and the external entities with which it interacts.

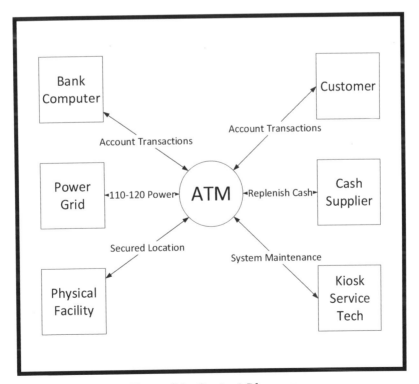

Figure 74 - Context Diagram

Control Chart

The Control Chart was invented by Walter Shewhart in 1924 and helps determine if a process being measured is In-Control or Out-of-Control. A process In-Control has no external factors impacting it, therefore its output variation is naturally random—as it should be. However, if a process is Out-of-Control, then some outside factors are impacting the primary process generating unwanted variation.

There are two ways to determine if a process is Out-of-Control. The first is to see if any data point is outside of the Upper Control Limit (UCL) or Lower Control Limit (LCL) of the analyzed data set.

The second method is to watch for seven consecutive data points in a row above, or below the center line (Mean), or cascading up or down. This second method is called the Rule of Seven.

The UCL and LCL are determined by measuring the average of the differences of the data, which is called the Standard Deviation, and then counting three standard deviations above and below the center of the data values (called the Mean of the data).

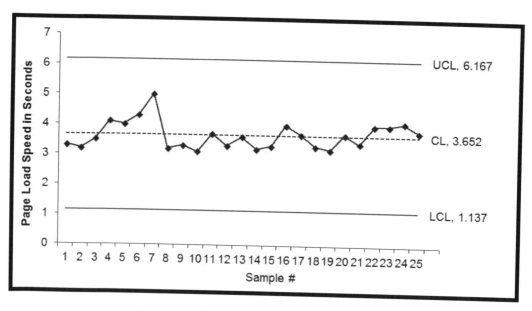

Figure 75 - Statistical Process Control Chart

Typically, a process is engineered so that its UCL and LCL are just within the customers specified variation range. We call this customer supplied range a Specification Limit, or a Tolerance.

A company trying to improve its page load speed could measure random samples of pages loading. The company might come up with a data set like the one shown in the figure above.

This chart reading indicates that the average page load speed is 3.652 seconds but can vary from 1.137 to 6.167 and still be In-Control. If suddenly we recorded a reading of 6.2 then some anomaly is responsible for that error and could be attributed to an operating system update, or database backup, or some similar action happening in the background that is competing for system resources.

Focus Group

A Focus Group is a small group of panelists who have experienced a problem or interacted with customers who have experienced a problem. Led by a facilitator, the panel answers questions collectively to educate the others in the meeting about a recurring problem. The purpose of a focus group is to learn more about a problem, not to solve it right there in the meeting.

Histogram

A Histogram is simply a chart showing a historical collection of data points concerning an output. Some Histograms compare multiple outputs to each other.

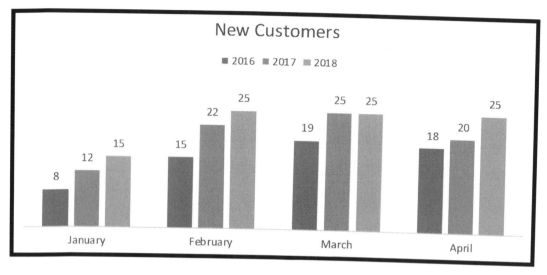

Figure 76 - Histogram

Pareto Chart

A Pareto Chart is used to show frequency of causes, prioritized by highest to lowest amounts.

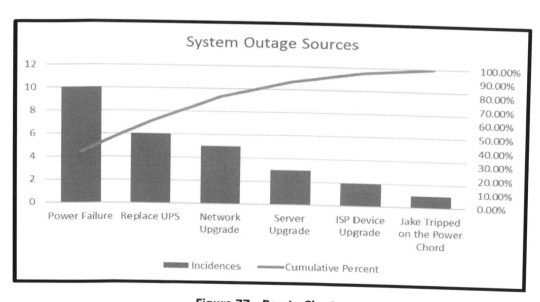

Figure 77 - Pareto Chart

The Bar Chart portion of the Pareto is linked to the left axis, showing frequency. The hairline portion is linked to the right axis, showing percentage. Then, for example, if you want to find the source of 80% of your system outages, you would see where the hairline meets the 80% indicator on the right axis and sum the bar frequencies left of that point.

In this example, the first three bars from the left represent 80% of the reasons why the system has gone down. The Pareto chart is ideal for tracking causes of defects and scrap.

Scatter Diagram

A Scatter Diagram is used to show if a relationship exists between two variables. The following figure is comparing a programmer's performance estimates versus actual performance. Over time a trend appears showing the actual time each work item takes which is about 1.5 times longer than the estimate. Therefore, in the future, the programmer should multiply his or her estimates by 1.5 to produce an estimate that is closer to reality.

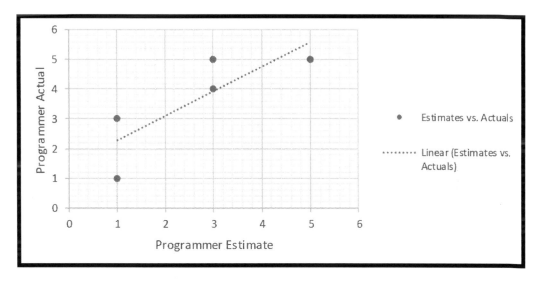

Figure 78 - Scatter Diagram

Sensitivity Analysis

A Sensitivity Analysis is a process where you first identify all the variables that impact an outcome, then one by one you change those variables and measure the result. This provides insight into which variables have the greatest impact on an outcome.

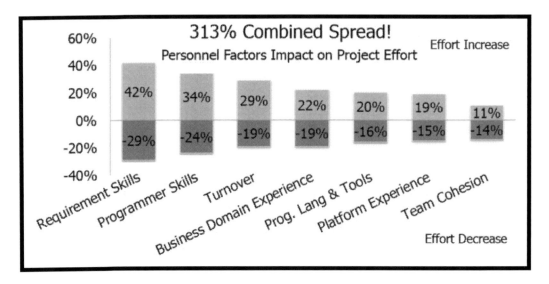

Figure 79 - Sensitivity Analysis

Data in the Figure above comes from Dr. Barry Boehm's COCOMO sensitivity analysis conducted in the early 80's. After analyzing hundreds of software projects his team determined that requirements skills have the greatest amount of impact on the overall effort required to produce software above any other factor listed. In other words, a trained requirements team would reduce the overall effort required to complete the software by 29%, whereas a poorly-trained requirements team would add 42% more effort than an average team to produce the same software.

SLAM Charts

SLAM Charts are tools for collecting performance data for process outputs. The simplest SLAM Chart, the first chart shown below, would indicate if a process completed successfully each day. A simple dot indicates success, and a number indicates a problem and corresponds to the problem explanation at the bottom of the chart.

The second example below is a more complex SLAM Chart showing statistical process control data for identifying trends and the amount of variation found in the system.

High Availability Systems

```
                              Day of Month Performance:
                              01 02 03 04 05 06 07 08 09 10
System                        11 12 13 14 15 16 17 18 19 20
                              21 22 23 24 25 26 27 28 29 30 31

                              .  .  .  .  .  .  .  .  .  .
Servicing ERP                 .  .  .  .  .  .  .  .  .  .
                              .  .  .  .  .  .  .  .  .  .  .

                              .  .  .  .  .  .  .  .  .  .
Servicing Backup              .  .  .  .  .  .  .  .  .  .
                              .  .  .  .  .  .  .  .  .  .  .

                              .  .  .  .  .  .  .  .  .  .
Code Library Backup           .  .  .  .  .  .  .  .  .  .
                              .  .  .  .  .  .  .  .  .  .  .

                              .  .  .  .  .  .  .  .  .  .
Daily Reconciliation Process  .  1  2  .  .  .  .  .  .  .
SLA: Complete before 12:00 pm .  .  .  .  .  .  .  .  .  .  .

                              .  .  .  .  .  .  .  .  .  .
Data Warehouse ETL            .  .  .  .  .  .  .  3  .  .
SLA: Complete every 20 minutes .  .  .  .  .  .  .  .  .  .  .
```

Notes:

1 - Process ran twice by mistake FIX: Added step to Kata Card.
2 - Cleanup from previous night's problems delayed process.
3 - Installed new code, process delayed replication for several intervals.
FIX: All new code installations will now be performed on weekends.

Figure 80 - Simple SLAM Chart

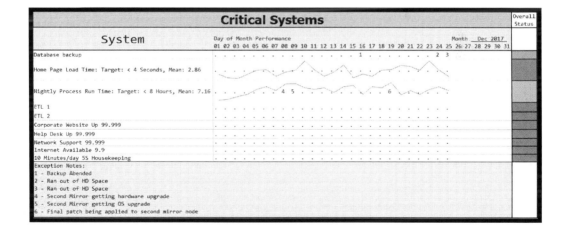

Figure 81 - Complex Slam Chart

Statistical Sampling

Statistical sampling is useful for identifying trends in a larger population, without having to collect information on the entire population. Thanks to the central limit theorem, patterns present in a large population will be observable in a smaller sampling of the same population.

Surveys are excellent tools for this but be sure your questions are unbiased and don't persuade the responder to select certain choices. Try to get at least 30 or more samples.

Suggestion Box

Be sure a suggestion box is visible and checked regularly. Provide contests for the number of suggestions that come from different teams. One simple litmus test of how well Stable is working in your environment is by measuring how many suggestions are made each month, quarter, and year.

Kata Card (Checklist, or Checksheet)

A Kata Card can be as simple as a checklist, but often it is more involved. Every widget moving through a production facility has information that needs to be tracked along with it. A Kata Card is used for this. They are also used to ensure the Process Owner has considered every risk factor known and ensured the problem is not present in this instance. In practice, they are sometimes called "checksheets," "travelers," or "routers."

Code Review Kata Card	
Item	**Pass/Fail**
Readable Variable Names	[ ]
No Loops >3 Deep	[ ]
App DateTime Sourced from the DataBase	[ ]

Figure 82 - Kata Card, or Checksheet

Data Analysis Tools

Data Analysis Tools are used for organizing information into contexts that facilitate discussions, generate knowledge, and support decisions.

Activity Network Diagram

An Activity Network Diagram is used to sequence activities, and then determine critical path, early start, late start, early finish, late finish, and total float time for each activity.

Activity Network Diagrams can be excellent tools for sequencing complex logic or logistical implementation procedures. A similar diagram that is used for the same purpose is called an Activity-On-Arrow diagram.

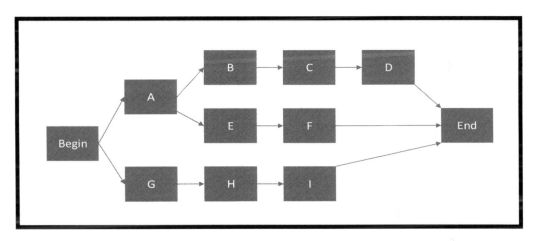

Figure 83 - Network Diagram

Affinity Diagram

An Affinity Diagram is a group brainstorming tool that collects ideas, opinions, and issues from a group and organizes them into topical clusters.

To perform the activity, each group member is given a stack of blank sticky-notes or index cards and encouraged to write a separate idea down on each card. After enough ideas are written, everyone's ideas are combined on a table-top or a wall, and a facilitator separates the ideas into topical clusters.

This tool is especially helpful for getting a lot of ideas or information from a group in a short amount of time, or if a group cannot agree on categories.

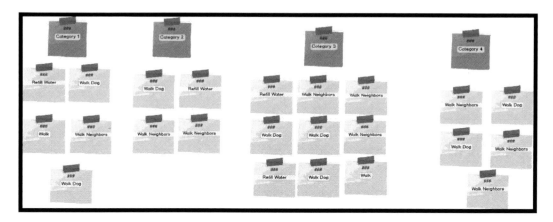

Figure 84 - Affinity Diagram

Burn-up Chart

A Burn-up Chart is an excellent tool for tracking team velocity over time. It's organized by units of time along the horizontal axis, and units of work along the vertical axis. Note the series along the top showing the total effort, or scope, identified and the total effort completed diagonal line near the diagonal straight line showing planned velocity.

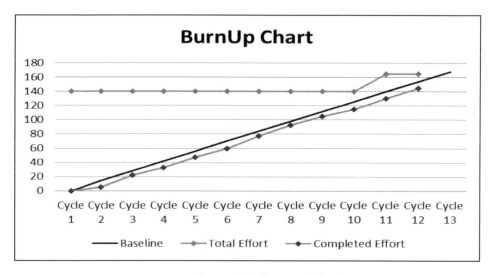

Figure 85 - Burn-up Chart

A Burn-down variant of this chart exists, but lacks the ability to show total effort scoped along the top.

Capability Chart

A Capability Chart is used to show various stages of capabilities concerning the relationship between two factors. The example below shows the range in proportional effort forecasted to generate thousands of source lines of code (KSLOC) over time. Note the tradeoff between overstaffing, understaffing, and optimal staffing. Also note the Productivity Index (PI) driving the forecasted output.

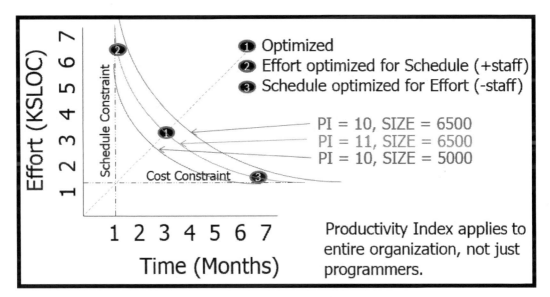

Figure 86 - Capability Chart

The chart above compares results of two different Productivity Indices (PI) on work to be performed. A team with a PI of 11 will complete the same amount of work in 3 months as a team with a PI of 10 will accomplish in 3.5 months. PI indexes can be calculated for teams using Lawrence Putnam's SLIM Method.

Cause and Effect Diagram, or Ishikawa Diagram

Also called a "Root Cause Diagram," "Ishikawa Diagram," or "Fishbone Diagram," this diagram was invented by a Japanese shipyard quality manager named Kaoru Ishikawa.

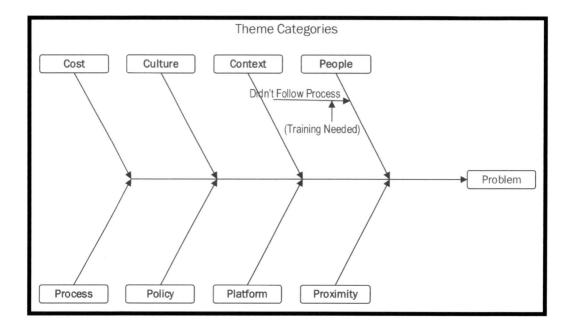

Figure 87 - Cause and Effect, or Ishikawa Diagram (Theme Based)

These diagrams are useful when a problem originates within a process and a group is collectively brainstorming the root cause. The Ishikawa Diagram is often combined with the 5 Why's questioning technique.

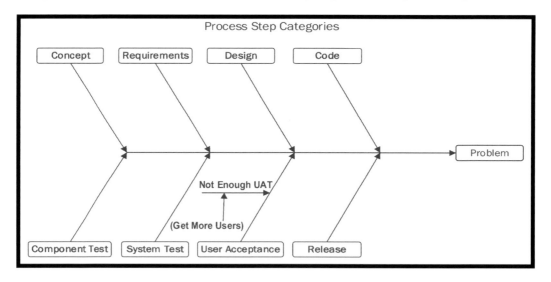

Figure 88 - Cause and Effect, or Ishikawa Diagram (Process Based)

There are two types of Ishikawa Diagrams. Theme-based diagrams, and process-based diagrams. Theme-based diagrams contain a collection of common areas for which a problem might occur. Process-based Ishikawa Diagrams contain all the steps of a process to be considered when searching for a root cause.

Facilitative Workshop

A Facilitative Workshop is a meeting called for many purposes. In it, a team can brainstorm for a solution, conduct an As-Is analysis, provide a To-Be walkthrough, or do any other group-interactive activity. Some Facilitative Workshops span multiple days.

Flowchart

Flowcharts are used to map out a process flow so that it can be analyzed, understood, and improved.

Modern flowcharts are typically organized by lines called swim lanes which indicate who is responsible for what activities.

The flowchart shown on the next page is a software testing process flowchart. Note the size difference with this diagram compared to the Sequence Diagram that follows it.

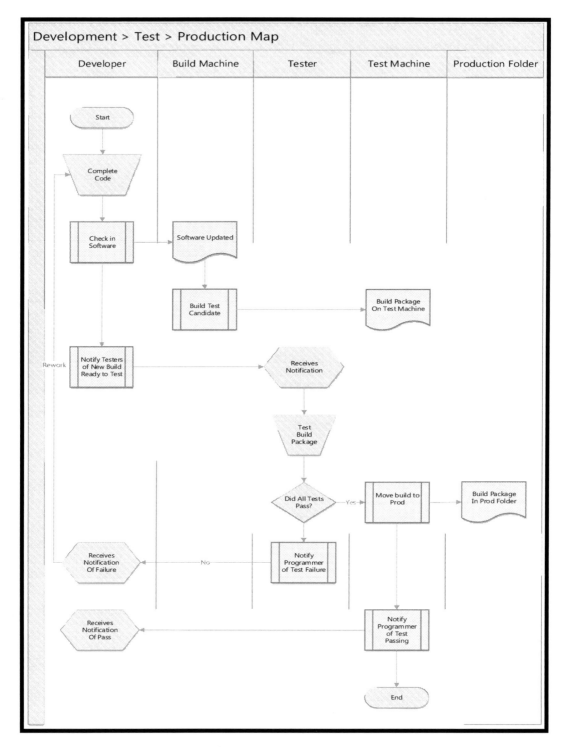

Figure 89 - Flowchart

This next figure is a UML Sequence Diagram, but it makes an excellent flowcharting tool showing process ownership and checkpoints. This diagram represents the same information as the flowchart on the previous page but is faster to build and uses a third of the space.

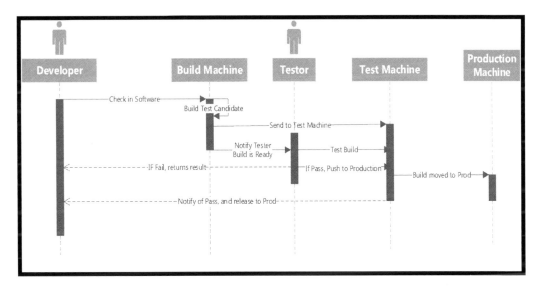

Figure 90 - UML Sequence Diagram used as a Flow Chart

Gantt Chart

A Gantt Chart is a common approach to laying out anticipated work in a dependent order. Gantt Charts are useful for showing project task breakdowns, dependencies, and to show the critical path through the workload, which helps keep the project on schedule.

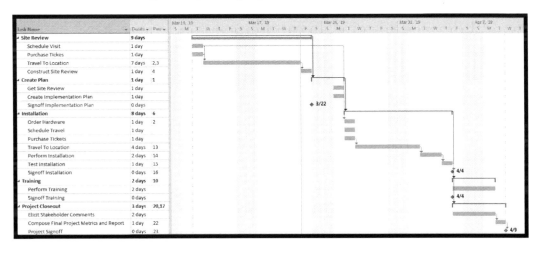

Figure 91 - Gantt Chart

Interrelationship Diagram

This tool is a way to show causes and effects within a scenario. A problem is written in a circle. Then one or more results of that problem are written in another circle, and an arrow is drawn to connect each cause and effect relationship. The process is then repeated on each one of the newly created circles until all cause and effect relationships are represented and connected.

A count is made of the incoming arrows versus the outgoing arrows for each circle, or node. Nodes with high incoming arrows represent the biggest areas of concern and are good targets for measurement. Nodes with the highest outgoing arrows represent the biggest sources of problems.

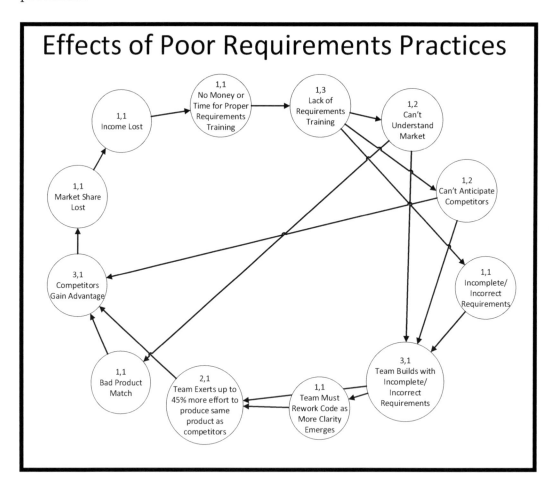

Figure 92 - Interrelationship Diagram

This is an excellent group collaboration tool. Many causes and effects can be exposed and discussed in a short amount of time with input from an entire team. Some people call this the "vicious cycle diagram."

Consider creating this diagram along with an Ishikawa diagram while flow-diagraming a system for improvement ideas. All three charts complement each other.

Matrix Diagram

A Matrix Diagram is used to organize and show relationships between factors. There are seven types of matrix diagrams: L, T, Y, X, C, R, and Roof shaped, depending on how many groups must be compared.

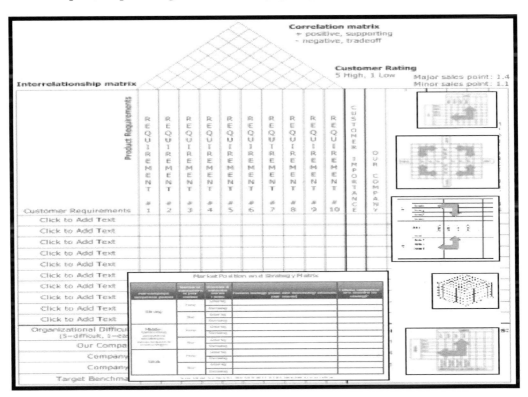

Figure 93 - Matrix Diagram

Prioritization Matrix

A Prioritization Matrix is a tool used to rank options against each other. When a Prioritization Matrix like the one shown below is populated (all blank cells contain 0 or 1) the relative value column on the right will show a list of projects prioritized from 10 down to 1.

Prioritization Matrix										
	Project 1	Project 2	Project 3	Project 4	Project 5	Project 6	Project 7	Project 8	Project 9	Project 10	Relative Value
Project 1	0	-	-	-	-	-	-	-	-	-	
Project 2		0	-	-	-	-	-	-	-	-	
Project 3			0	-	-	-	-	-	-	-	
Project 4				0	-	-	-	-	-	-	
Project 5					0	-	-	-	-	-	
Project 6						0	-	-	-	-	
Project 7							0	-	-	-	
Project 8								0	-	-	
Project 9									0	-	
Project 10										0	
If project in Row is more desirable than project in Column enter a '1'. If not, enter a '0'. The project with the greatest relative value is weighted highest.										

Figure 94 - Prioritization Matrix

Process Decision Program Chart (PDPC)

A Process Decision Program Chart is used to pre-plan risk contingencies. Any anticipated risk can have a plan made to mitigate the problems posed by the risk.

PDPC Charts are excellent compliments to addressing identified risks uncovered during a risk analysis and are often used with a Failure Mode Effects Analysis (FMEA).

The mark of seasoned professionals in any part of IT is a healthy respect for risk management and their generation of plans to anticipate and mitigate risks.

If you think about it, the reward for completing a large and risky project successfully is a larger and even riskier project. This continues throughout an IT professional's career until risk management becomes a time-consuming and serious part of planning and reporting. It's best to learn how to do it properly now.

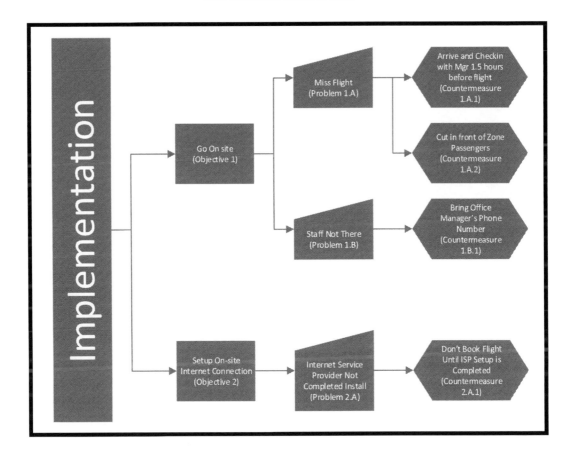

Figure 95 - Process Decision Program (PDPC) Chart

Radar Chart

A Radar Chart is useful for identifying bottlenecks or problem areas that need improvement.

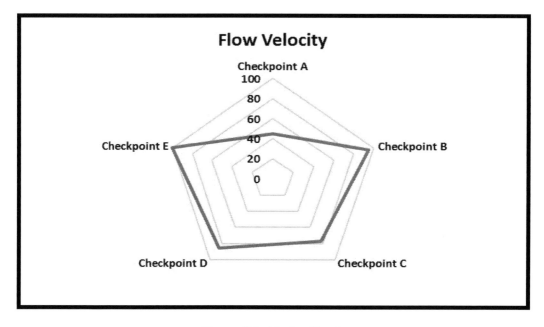

Figure 96 - Radar Chart

Tree Diagram

A Tree diagram is used to examine two or more variables. The first variable is represented at the first split. The second variable is represented at the second split, and so on. The result is that all possible resulting combinations are represented at the end (rightmost) nodes.

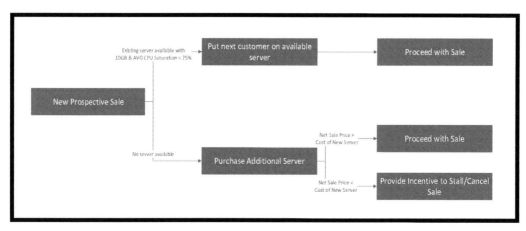

Figure 97 - Tree Diagram

Glossary of Terms

5S. A workplace organization method that uses a list of five phrases, all beginning with "S": Sort, Set in order, Shine, Standardize, Sustain. Adapted from Japanese: Seiri, Seiton, Seiso, Seiketsu, Shitsuke.

Acceptance Screen. A mechanism used to mistake-proof an incoming or outgoing asset so that it will interact correctly with its intended environment. In Japanese this is called "Poka-Yoke."

After Action Review (AAR). A meeting called right after an event where the facilitator presents what happened compared to what was supposed to happen, and the group discusses why it happened the way it did, if it could have happened better, and how it could have happened better.

Alert System. A system put in place to show the real-time status of assets in an environment. This solution could be a scoreboard monitor hanging in a work area, or a message emailed or texted to a group list, or something similar.

Apollo Team. Also called the Apollo Syndrome, a phenomenon identified by Dr. Meredith Belbin where teams of highly capable individuals can collectively perform badly. This is attributed to excessive time in destructive debates, omitting needed work, working in silos, and avoiding necessary confrontation.

Asset Controls. All the governing information and instructions pertaining to an asset. This includes Asset Recovery Models, Process Asset Matrix, Asset Maintenance Logs, and Asset Purchase and Warrantee Information.

Asset Recovery Model (ARM Sheet). A guide associated with an asset, for quickly diagnosing and remedying one or more problems that may occur while using the asset. The model is based on past problems and it gets added to as new problems and solutions are discovered. Its purpose is to shorten asset downtime.

CCAPA Committee. A committee staffed to address nonconforming problems discovered by customers while using the product or service. CCAPA means Correction, Corrective Action, and Preventative Action, which is a pattern for dealing with problems. Correction is fixing the specific customers situation. Corrective Action means changing production so that no one else experiences that problem in the future. Preventative Action means changing any other similar processes that may benefit from a similar Corrective Action. A CCAPA Committee should meet daily, weekly, or once each Master Cycle.

Checkpoint. A coordination point between processes or process steps where a completed Kata Card is handed off to the next Process Owner at the receiving process step. At the final checkpoint, the accumulated stack of Kata Cards is given to the Master Chief.

Coaching Kata. A pattern used by a Master Chief to motivate a Process Owner, in a structured manner, towards a goal.

Configuration Management System (CMS). A storage repository for Master Lists, Process Controls, and Asset Controls.

Control Points. A strategic checkpoint measured to provide summary information for the organization.

Crux Design. An analysis, design, and estimation technique used to quickly model a reasonable representation of a system and provide a corresponding estimate for time and resources needed to construct it.

Culture. The anticipated interactions between two or more people. In organizations, culture is typically polarized on two axes: Independent vs. Team-focused, and Flexible vs. Rigid. Culture is ever-changing and can be altered over time through shared memories and repeatedly emphasizing important topics in conversations.

Cumulative System Performance Chart. A chart showing Master Cycle performance trends over time, to be updated at the end of every Master Cycle. Typically, this chart or a variation of it is brought to a Quarterly Business Review (QBR) meeting.

Customer Value. Utility a paying customer receives in exchange for currency.

Effort Point. A reasonable amount of work that can be accomplished by one skilled in the art, with minimal distraction. Sometimes called a 'Stone.'

Facilitative Workshop. A group-based meeting held for many purposes, namely brainstorming, knowledge-transfer, walk-throughs, or other planning discussions. Some Facilitative Workshops can span several days.

Failure Modes and Effects Analysis (FMEA). A systematic method for evaluating a process to identify where and how it might fail, and to assess the relative impact of different failures, in order to identify parts of the process most in need of change or contingency planning.

Focus Group Meeting. A meeting held with a small group of people familiar with an issue to understand the issue better.

Hidden Factory. Excess work, rework, scrap, wait-time, gold-plating and other cost-draining practices that represent effort expended that the customer does not need to pay for.

Hot-fix Haze. Any environment without a functioning quality program, characterized by frequent rework, fire-fighting, and heroic efforts to deliver value.

Initial Quality Assessment (IQA). Inspecting incoming materials and information to be sure they meet requirements before incorporating them into an internal product build.

Institutional Knowledge. Information learned on the job that has been transferred to standard operating procedures, lessons learned, risk lists, asset recovery models, and other permanent records so that the entire company will benefit from what was learned. Continuous Improvement is based on adding to and revising Institutional Knowledge.

Just-In-Time Manufacturing. An industrial technique where external material suppliers are challenged to deliver needed goods right as they are needed for production, and not sooner. This eliminates expensive warehousing costs.

Kaizen Blitz. A team improvement effort organized by a request for help from a Master Chief or a Process Owner. A Kaizen Blitz is different from a Kaizen Project in that a Blitz is triggered as needed and should only last a few days, whereas a Kaizen Project is pre-planned and may consume an entire Master Cycle.

Kaizen Project. A problem within the department, or with a product or service, that the team or everyone on the team takes on to improve during the Master Cycle. This challenge is identified during the Cycle Retrospective Meeting and reported on during the next Cycle Review

Meeting.

Kaizen Team. A team created as needed to solve a problem. These teams seem to produce the best results if they are staffed by volunteer participants. Sometimes these are called Ad-hoc Kaizen Teams, or Ad-hoc Teams.

Lessons Learned Log. A historical record of knowledge gained over time. Typically, this is part of a Process or Asset Recovery Model, and critical findings to watch out for in the future are added to the Kata Card for that step.

Market Summary Report. A summary of happenings or news from the customer base delivered every morning by the Master Chief to the Process Owners in the morning Kaizen Stand-up Meeting. This information is sourced from senior management, operations, and the Service Desk.

Master Cycle. A cadence rhythm established for reporting and measuring purposes. It should last between one to four weeks. If your group works alongside an Agile group, you should synchronize your cycles with theirs so that you can report together.

Master Kaizen. In a large Stable Framework™ implementation, it may be necessary to stagger Kaizen Meetings. The Master Kaizen Meeting is a meeting of representatives from each Kaizen Meeting who report on the highlights of what their teams are doing.

Master Lists. Lists of important information needed to be maintained within a system.

Net Promoter Score. A score measuring how probable a customer base is to recommend a business.

Nonconforming Issues. Problems found by customers and brought to the attention of the Process Owner, usually via the Service Desk. These problems are prioritized and addressed by the Process Owner, or the CCAPA Committee.

One-piece Flow. Also called One-piece flow. The idea that producing one item at a time provides value to the marketplace faster than batching items through production. Early items within the batch must wait for all the other items to be completed in the batch before they can all move to production. For this reason, we minimize task switching and minimize work-in-process as much as possible.

Operational Excellence. A standard of execution and reporting so transparent that senior management or even a stranger off the street can look at a performance console and understand everything important going on within the department. This frees senior management's time so they can focus on growing the business.

Performance Console. Sometimes called an Information Radiator, this is a dedicated space on a wall or rolling board for the team to report on the performance of their processes, customers, and supplier services.

Personal Coefficient. A factor introduced by Fredrick Taylor in his book *Scientific Management.* which attributes differences between separate individuals performing the same task.

Personas. Fabricated personalities created by the team and used to calibrate user experience expectations and gain insights into additional user needs.

Plan-Do-Check-Act. A hypothesis testing pattern created by Walter Shewhart and made famous by W. Edwards Deming. When you have an idea you "Plan" how to try it, "Do" the experiment, "Check" the results, and then "Act" by changing production if the result was desirable.

Planning Package. A vague unit of work within a Work Breakdown Structure (WBS) that has not yet been decomposed into one or more work packages, due to excessive remaining ambiguity.

Poka-Yoke. A Japanese word that means mistake-proofing, or error prevention. This is a technique where the use of a design change or added tool will prevent inadvertent human error. For example, the plastic shim in a USB port and plug, that prevents the user from inserting the plug the wrong way is a Poka-Yoke. In Stable we call these Poka-Yokes "Acceptance Screens."

Process Controls. All the information and instructions governing the use, maintenance, and recovery of a process. This includes Standard Operating Procedures, Process checklists we call Kata Cards, Service Levels, Recovery Models, etc.

Process Kata. When a Process Owner executes a repeatable process, this is called a Process Kata. A certificate indicating it was completed successfully is the completed, signed Process Kata Card.

Process Quality. How well the processes were performed, according to the written procedure and with the least amount of excess, waste, rework, delays, or any other unnecessary byproduct. This is measured during Quality Assurance efforts.

Product Quality. How well the product matches the expectations of its customers. This is measured during Quality Control efforts.

Product Backlog. A list containing intentional functionality to add to a project workload.

Pull System. A manufacturing system triggered to produce the next product when the next sale occurs. This configuration is the most efficient due to minimal warehousing needs and assumes all upstream materials and components are in place queued for the next sale.

Push System. A manufacturing system based on queuing batches of products and storing them until the next sales are made. This system is less efficient than a pull system.

Quality Debt. The fact that costs involved with repairing a defect rise almost exponentially as they cascade down from where they were introduced to later value stream steps.

Quality Planning. The initial setup and identification of quality goals within an environment, or a project. In the Stable Framework™ this involves establishing your Configuration Management System, Master Services, Supplier Services, Service Levels, Standard Operating Procedures, Kata Cards, Asset Recovery Models, Work Queues, Kanban Board, Suggestion box, Master Cycle, Cycle Planning Meeting, Daily Kaizen Meeting, Cycle Review, and Cycle Retrospective Meetings.

Requirements Rot. A natural phenomenon where changes in the environment, marketplace, government regulations, and human interests lead to requirements becoming less relevant over time if neglected.

Service Level Agreement (SLA). A formal or informal contract between a customer and a supplier for a certain level of service to be expected.

Service Level Attainment Monitoring (SLAM) Chart. A chart showing the Master Cycle performance of a process of interest, against its Service

Level Agreement, or Service Level Goal.

Service Level Goal (SLG). An informal standard internally set by the production team to provide services at a particular level.

Service Level Histogram. A histogram displaying the past trend of service level results.

Service Oriented Architecture. A style of software design where services are provided to other components across the enterprise, via a common documented protocol.

Single-Piece Flow. See One-Piece Flow.

Standardize-Do-Check-Act. A variation on the Plan-Do-Check-Act model used for Quality Planning while creating Standard Operating Procedures (SOP's). The Process Owner "Standardizes," or creates the SOP, "Does" the process step, "Checks" to be sure the SOP matches the step, and if not "Acts" by updating the written SOP, until it is accurate enough that somebody skilled in the art, but not necessarily from that environment could repeat the steps with that level of written instruction.

Stone. A reasonable amount of work that can be accomplished by one skilled in the art, with minimal distraction. Another name for Effort Point.

Supply Chain. The collection of upstream suppliers providing needed technologies, information, supplies, or services as inputs to your processes. In the Stable Framework™, we call these Supplier Services.

T.I.M.W.O.O.D.S. An acronym to help remember the eight wastes in Lean Manufacturing Theory. They are: Transport, Inventory, Motion, Waiting, Over Production, Over Processing, Defects, and Skills Underutilized.

Takt Time. This is the speed at which your customers consume your products. Used as a calibration mechanism for performance tuning the rate at which you should be supplying services.

Team Development Model. A pattern identified by Bruce Tuckman describing how teams Form, Storm, Norm, Perform, and then Adjourn through natural transitions.

Tribal Knowledge. Knowledge developed while on the job and stored in the minds of workers. The goal of Process Owners is to systematize their knowledge so that it is transferred into Institutional Knowledge. Once this transition occurs, real Process Improvement can begin.

Trigger. The starting mechanism to begin a process.

Value Proposition. The incentive that creates a customer. Each process should have a value proposition.

Value Stream. The collection of upstream and downstream processes, process steps, materials, information, and suppliers producing a product or service of interest to a customer.

Victim Triangle. Also called the Drama Triangle, a model of human dysfunctional interaction where an underperforming "victim" complains about a "persecutor" disabling the "victim's" ability to perform an agreed-upon task and asks the manager to "rescue" the "victim" from the consequences of a neglected task by excusing it this time. Instead, the manager should challenge the "victim" to express what he or she could have done differently under the same oppressive circumstances to still meet the commitment, and then challenge the "victim" to do that next time.

Work Breakdown Structure (WBS). The total scope of a project to be delivered, organized into logical deliverable structures called Work Packages, or Planning Packages.

Work In Progress (WIP). Also called Work in Process, this term describes all the work presently being performed by the team, or an individual on the team. WIP-Limits are established to ensure the system is streamlined, but not overburdened. Overburdening a system causes unnecessary delays due to task switching by humans and equipment.

Work Package. The subcomponent within a Work Breakdown Structure to be completed or delivered, usually composed of one or more activities.

WIP-Limit. The amount of bandwidth available or dedicated to a type of work. If a WIP-Limit is breached, a bottle-neck emerges inhibiting the flow of value through the system.

Bibliography

Chapter 1 – The Problem

- Dr. Atul Gawande; *The Checklist Manifesto: How to Get Things Right*; Jan 2011; Picador
- The Standish Group International, http://www.StandishGroup.com
- Gordon Training International, http:// http://www.gordontraining.com/free-workplace-articles/learning-a-new-skill-is-easier-said-than-done/#

Chapter 2 – The Hidden Factory in IT

Chapter 3 – System Thinking and Value Streams

- Mark Schwartz; *The ART of BUSINESS VALUE*; 2016; IT Revolution

Chapter 4 – Six Forms of Process Improvement

- John L. Lee; *Rising Above It All*; Oct 2012; iUniverse
- Dean R. Spitzer; *Transforming Performance Measurement*; 2007 Amacom
- Douglas W. Hubbard; *How To Measure Anything*; 2010; Wiley & Associates
- John Doerr; *Measure What Matters*; 2008; Bennett Group, LLC

Chapter 5 – The Components of a Quality Program

- W. Edwards Deming; *Out Of The Crisis*; Aug 2000; MIT Press
- J.D. Power Ratings; *Korean Brands Lead Industry in Initial Quality, While Japanese Brands Struggle to Keep Up with Pace of Improvement*, http://www.jdpower.com/press-releases/2015-us-initial-quality-study-iqs

Chapter 6 – Common Quality Programs

Chapter 7 – The Stable Framework™

- Atul Gawande; *The Checklist Manifesto*; Jan 2011; Metropolitan Books
- Kevin J. Duggan; *Design for Operational Excellence: A Breakthrough Strategy for Business Growth*; Sep 2011; McGraw-Hill

- Christian Moore; *The Resilience Breakthrough: 27 Tools for Turning Adversity into Action*; July 2014; Green Book Group Press
- Jack Welch and Suzy Welch; *Winning: The Ultimate Business How-To Book*; Oct 2009; HarperCollins Publishers

Chapter 14 – Committees, Groups, and Kaizen Events

- Masaaki Ima; *Gemba Kaizen: A Commonsense Approach to a Continuous Improvement Strategy*; Jun 2012; McGraw-Hill

Chapter 15 – The Stable, or "Hybrid" Project

- PMI-ACP®, PMI®, PMP® are registered trademarks of the Project Management Institute

Chapter 16 – Using Stable in Operations

Chapter 17 – Using Stable in Implementation

Chapter 18 – Using Stable in DevOps

- Gene Kim, George Spafford, and Kevin Behr; *The Phoenix Project: A Novel about IT, DevOps, and Helping Your Business Win*; Jan 2013; It Revolution Press
- Gene Kim, Jez Humble, Patrick Debois, and John Wellis; *The DevOps Handbook: How to Create World-Class Agility, Reliability, & Security in Technology Organizations*; 2016; IT Revolution Press

Chapter 19 – Using Stable in Product Development

Chapter 20 – Using Stable with Agile Development

- PMI-ACP®, PMI®, PMP® are registered trademarks of the Project Management Institute

Chapter 21 – Stable Portfolio Management

- International Software Benchmarking Standards Group, http://www.ISBSG.org

Chapter 22 – Getting Started with Stable

Chapter 23 – 20 Data Collection & Analysis Tools

- Timothy J. Clark; *Success Through Quality: Support Guide for the Journey to Continuous Improvement*; Feb 1999; ASQ Press
- Lawrence H. Putnam and Ware Myers; *Measures for Excellence: Reliable Software On Time, Within Budget*: Oct 1991;

Index

D

E

N

O

P

CERTIFICATIONS ARE AVAILABLE TO ENSURE THE INTEGRITY OF THE FRAMEWORK AND THE EFFECTIVENESS OF RESULTS WHEN PRACTICED.

FOR MORE INFORMATION:

WWW.STABLEFRAMEWORK.ORG

MICHAEL J. BERRY

CSMC, CSPRO, PMP, PMI-PBA, PMI-ACP,
ITIL, CSM, CSPO, PSM I, PSPO I, ISTQB-CTFL

Michael J. Berry has worked in the software industry since 1987. During this time he has developed enterprise software for banks used in over 2500 branch sites in six states, and led teams that have developed and supported manufacturing software used on five continents, and medical software used in over 600 clinics. He has consulted with and trained thousands of technology professionals during the past ten years.

Michael is an adjunct professor at the University of Utah and Utah Valley University. He is an IEEE Senior Member, a John C. Maxwell leadership coach, and the executive producer of SoftwareDevelopmentDays.com. In 2007 he founded Red Rock Research, a professional development organization. Michael holds a B.S. in Information Systems and Technologies from Weber State University.

Michael can be reached at Michael@StableFramework.org

74869446R00152

Made in the USA
Columbia, SC
14 September 2019